Introduction to Octave

For Engineers and Scientists

Sandeep Nagar

Apress®

Introduction to Octave: For Engineers and Scientists

Sandeep Nagar
New York, USA

ISBN-13 (pbk): 978-1-4842-3200-2 ISBN-13 (electronic): 978-1-4842-3201-9
https://doi.org/10.1007/978-1-4842-3201-9

Library of Congress Control Number: 2017960430

Cover image by Freepik (www.freepik.com)

Managing Director: Welmoed Spahr
Editorial Director: Todd Green
Acquisitions Editor: Steve Anglin
Development Editor: Matthew Moodie
Technical Reviewer: Massimo Nardone
Coordinating Editor: Mark Powers
Copy Editor: Kezia Endsley

Distributed to the book trade worldwide by Springer Science+Business Media New York, 233 Spring Street, 6th Floor, New York, NY 10013. Phone 1-800-SPRINGER, fax (201) 348-4505, e-mail orders-ny@springer-sbm.com, or visit www.springeronline.com. Apress Media, LLC is a California LLC and the sole member (owner) is Springer Science + Business Media Finance Inc (SSBM Finance Inc). SSBM Finance Inc is a **Delaware** corporation.

For information on translations, please e-mail rights@apress.com, or visit http://www.apress.com/rights-permissions.

Apress titles may be purchased in bulk for academic, corporate, or promotional use. eBook versions and licenses are also available for most titles. For more information, reference our Print and eBook Bulk Sales web page at http://www.apress.com/bulk-sales.

Any source code or other supplementary material referenced by the author in this book is available to readers on GitHub via the book's product page, located at www.apress.com/9781484232002. For more detailed information, please visit http://www.apress.com/source-code.

Printed on acid-free paper

Dedicated to my wife Rashmi and my daughter Aliya

Table of Contents

About the Author

Sandeep Nagar, PhD (Material Science. KTH, Sweden) teaches and consults on the use of Octave and other open source software. In addition to teaching at universities, he frequently gives workshops covering open source software and is interested in developing hardware for scientific experiments.

About the Technical Reviewer

Massimo Nardone has more than 22 years of experience in security, web/mobile development, cloud, and IT architecture. His true IT passions are security and Android.

He has been programming and teaching others how to program with Android, Perl, PHP, Java, VB, Python, C/C++, and MySQL for more than 20 years.

He holds a Master of Science degree in Computing Science from the University of Salerno, Italy.

He has worked as a project manager, software engineer, research engineer, chief security architect, information security manager, PCI/SCADA auditor, and senior lead IT security/cloud/SCADA architect for many years.

Massimo's technical skills include security, Android, cloud, Java, MySQL, Drupal, Cobol, Perl, web and mobile development, MongoDB, D3, Joomla, Couchbase, C/C++, WebGL, Python, Pro Rails, Django CMS, Jekyll, and Scratch.

He currently works as the Chief Information Security Office (CISO) for Cargotec Oyj.

He was a visiting lecturer and supervisor for exercises at the Networking Laboratory of the Helsinki University of Technology (Aalto University). He holds four international patents (in the PKI, SIP, SAML, and proxy areas).

Massimo has reviewed more than 40 IT books for different publishing companies and he is the co-author of *Pro Android Games* (Apress, 2015).

Acknowledgments

I wish to thank Steve, Mark, and the whole team at Apress to bringing this book to fruition. I also wish to thank the Octave community for answering questions on forums, which helped me learn difficult concepts with ease.

CHAPTER 1

Introduction to Octave

1.1 Introduction to Numerical Computing

Modern times have seen an exponential growth in scientific knowledge. Computing devices have benefited the most from this increase in knowledge. They started with mechanical solutions, whereby lever- and pulley-based computers were used to perform complex calculations with progressively fewer interventions. But mechanical systems were notoriously slow and inefficient and there were energy concerns too. When vacuum tube based transistors were invented, they were applied to this domain. Vacuum tube based computers entered the research labs of academia and industry alike. During WWII, they were used to run programs to crack enemy codes as well as to simulate various scenarios for designing weapons. This infused much needed money and talent in this area and, within a few years, the overall efficiency of computing devices saw an amazing exponential increase.

Simulating a real-world phenomenon involves solving equations governing these issues. Numerical simulation involves defining the problem for a digital computer. This can be achieved in two ways:

- Using a programming language and encoding every step used for numerical computation
- Using a specialized software framework that presents a general framework to define a mathematical problem that the computer understands

S. Nagar, *Introduction to Octave*, https://doi.org/10.1007/978-1-4842-3201-9_1

Scientific computation was initially performed by the first method. Programming languages like FORTRAN, C, and C++ became very popular. Even today these languages, and recent ones like Python and Julia, are still widely used for this purpose. But during the same time, the need for a specialized numerical computing framework was also recognized. Using programming languages, you could make a generalized scheme for numerical computing in which you could define a scientific problem. One of the biggest advantages of such an approach is that you can define a lot of library functions that can be simply used as and when required, instead of each user writing them down each time for a different problem. Hence, over the period of time a number of software programs came into existence.

One of them was MATLAB, and it became very popular all over the world. It is sometimes called the "language of engineering" for the right reasons, since most engineering problems can be easily defined using it. Engineers can concentrate on defining a problem rather than writing efficient code (which they can simply pick from a library). Being commercial software, MATLAB comes with a price as well as with a restrictive license. With the introduction of open source licensing, there was a need for an open source alternative. This is where Octave came into being. This book presents usage of Octave as an effective alternative to MATLAB.

1.2 Analytical vs. Numerical Schemes

Analytical schemes to solve mathematical problems involve deriving equations describing a system using relationships between various parameters and then solving these equations by either using invented functions (mapping of variables from one domain to another) or inventing new functions that fit the purpose. On the other hand, a numerical scheme also requires describing the system using a relationship between various parameters and functions, but deriving a solution has a marked difference. This can be shown with a simple example.

Let's try to find a value of x that can satisfy the equation $f(x) = x + 25$. Finding an analytical solution involves the following steps:

1. $x + 2 = 0$
2. Add −2 to both sides
3. $x = -2$ is the answer

On the other hand, a numerical solution using the *bisection method* involves first guessing a value as a solution of the equation and then following the scheme as shown here:

1. Let's guess 3 as the answer
2. $f(3) = 1 + 2 = 3 > 0$ so let's guess $f(-3) = -3 + 2 = -1 < 0$
3. Since 3 results in an answer more than 0 and −3 results in an answer less than 0, an average value is calculated for both as follows:

   $$\frac{3+(-3)}{2}=0$$

   The $f(0)$ is calculated and replaced with the same initial guess, depending on if the result is less than or greater than zero.

4. These steps are repeated successively until we reach the true value, i.e., −2 (as found using the analytical solution).

Now the question arises that if we have analytical solutions, why should we even care for finding numerical solutions? The main reason is that sometimes we don't have an analytical solution. Try to solve $e^{x.sin}\left(x^2 - tan^{2/3}(x) - 4x^3 + 2x^2 - 4 = 0\right)$. Finding solutions would require too much human effort (may even take more than a lifetime in some cases). Moreover, complex problems involving advanced structures like differentiations, integrations, etc., are very difficult to solve using analytical solutions. For these purposes, numerical schemes have been defined.

As time progressed, various schemes to define analytical functions like differentiation, integration, trigonometric, etc., were written for digital computers. This involved their digitization, which certainly introduces some errors. Knowledge of error introduced and its proper nullification could yield valuable information quicker than using analytical results. Thus, it became one of the most actively researched fields of science and continues to be one. The search for faster and more accurate algorithms continues to drive innovation in the field of numerical computing and enables humanity to simulate otherwise impossible tasks.

1.3 Tools for Numerical Computation

While all problems can be coded in programming languages, we need to change the approach to computing, file management, etc. when we change the microprocessor platform, operating system, or both. This hinders interoperability. Modern programming languages address some of these issues but the need for specialized software for numerical computing, where predefined tools can be simply *called* as and when required, was being felt in academia. A number of attempts were made in this direction.

A number of alternatives exist to perform numerical computations. Programming languages written to handle mathematical functions like FORTRAN, C, Python, and Java, to name a few, can be used to write algorithms for numerical computation. Specialized software packages like MATLAB, Scilab, and Mathematica also provide specialized solutions to particular fields of problems. Their rich libraries now run in many GBs of data.

Among them, MATLAB became tremendously popular among the scientific community starting in 1984. The cheap availability of digital computing resources propelled its use in industry and academia to such an extent that virtually every lab needed MATLAB. It was embraced by academia as well as industry and in some cases, became a standard tool for computational work. An engineer who was not trained on MATLAB was less employable than others and, hence, a lot of universities adopted it in their curriculum.

But MATLAB has two serious drawbacks: its price and its licensing requirements. It started with a set of freely available tools written by academics but, when it became a commercial product, it came with a commercially restrictive license and a hefty price tag. It was the license that troubled academics more than the cost, because the license blocked sharing the software and even required an additional cost to do research work apart from teaching MATLAB.

It wasn't cost-prohibitive for well funded western universities, but it proved to a costly piece of software for rest of the world, particularly for third-world countries. This part of the world, which has an otherwise large scientific community, needed an open sourced alternative to MATLAB. Thus Octave and Scilab came into existence.

Whereas Scilab is extremely powerful, it was not compatible with MATLAB syntax-wise. It derived its origin from the same pieces of code from which MATLAB was born, but it split the other way and defined different types of files for computations and improved on the syntax accordingly. Hence, you cannot run MATLAB directly using Scilab. On the other hand, Octave was developed so that MATLAB `.m` files could directly run on Octave software.

1.4 A Brief History of Octave

MATLAB was developed by Cleve Moler [1], who was a math professor at the University of New Mexico, teaching numerical analysis and matrix theory. As a PhD student, he initially wrote a lot of code in FORTRAN to solve systems of simultaneous linear equations involving matrix algebra. He ultimately gave this the name MATrixLABoratory (*MATLAB*). As a professor, he wanted his students to be able to use their new packages without writing FORTRAN programs. Hence, in late 1970s, the first version of MATLAB came out (written in FORTRAN). There were 80 functions for performing calculations involving linear algebra problems. Further down the line, Jack Little and Steve Bangert reprogrammed MATLAB in C with additional features for producing a commercial version of the software. Together, all three of them founded The MathWorks [2] in California in 1984, which develops, maintains, and distributes MATLAB and its products worldwide. MATLAB as proven to be an excellent tool for numerical methods [3].

So many tools and features have been added to the base package of MATLAB that along with a rich set of libraries, the installation requirements run in many GBs of data. MATLAB became tremendously popular within the scientific community. It is being used by more than 5000 universities worldwide. It is sometimes rightly termed the "language of engineering". Cheap availability of digital computing resources propelled its usage in industry and academia to such an extent that virtually every lab needs MATLAB.

Octave, on the other hand, was conceived in 1988 [4]. Initially, it was conceived to be merely companion software by James B. Rawlings [5] of the University of Wisconsin-Madison and John G. Ekerdt [6] of the University of Texas, for an undergraduate-level textbook on chemical reactor design. They realized that chemical engineering students were finding it difficult to code in FORTRAN. Instead, they wanted a solution where students could concentrate on solving chemical engineering problems. So they conceived a solution where they could use an *interactive environment* like Octave so that students could learn quickly and start coding in a few hours. This was a similar solution to the one provided by Cleve Moler and which ultimately became MATLAB.

For the next five years, development proceeded toward making Octave almost as good as basic MATLAB. On February 17, 1993, Version 1.0 was released. Contrary to popular belief, the name Octave is not related to music. It was actually named after Dr. Octave Levenspiel [7]. He was a former professor who wrote a famous textbook on chemical reaction engineering and was also well known for his ability to do quick, "back of the envelope" calculations.

Octave is shared under a GNU General Public license [8] and hence it is free to modify and redistribute as defined by the license. Being open sourced, it grew rapidly as one of the hottest open source projects, where developers (mostly students) from all over the world contributed code to the project. This enriched the main program as well as various specialized packages. Its large base of library functions makes it an obvious choice for defining engineering problems. Since it can run MATLAB files without any major changes, it became popular with students, as they could install Octave on their personal computers and study at home too.

This book introduces Octave for absolute beginners. Even if you have never used MATLAB, you can start with Octave. But over time, you are encouraged to become a developer yourself. Developers enrich the library functions and share within the community. The community of users and fellow developers test and report bugs. These are then rectified in a collaborative manner. This ecosystem of collaborative development is the backbone of open source scientific computing. Users may find more information about Octave development at [9]. As an example, check out a host of community developed packages, which are listed in reference [10].

1.5 Octave vs. Other Alternatives

Octave is an open source alternative that can run MATLAB code. Existing MATLAB users can swiftly change to this new system. Similarly, new users can learn to code in Octave and then shift to a MATLAB environment as and when required. GNU Octave version 4.0.3 presents a Graphic User Interface (GUI) too, which proves to be an easier option for beginners. For this reason, it has been chosen for this book. The book's code will also run older and future versions well, provided that future versions choose to remain compatible with the present version.

Other alternatives include Scilab and programming languages like Python, C, C++, and Java. They have their own merits and drawbacks and you are advised to decide based on your particular needs. Octave is a good choice for prototyping the problem quickly and checking the results.

These alternatives are better when you are working with web-based data collection, analysis, and visualization. Octave is a high-level language, primarily intended for numerical computations. Octave has a rich library of tools for solving numerical linear algebra problems, finding the roots of non-linear equations, integrating ordinary functions, manipulating polynomials, and solving ordinary and partial differential and differential-algebraic equations. This makes it suitable for most of the basic numerical computational work. When you are concerned about speed and need multi-core programming for data distributed over multiple web servers, you might opt for coding in specific programming languages written for high-performance computing like Julia and numpy/scipy (Python), or simply C/C++. But you can still prototype parts of such problems in Octave for simplicity of understanding.

1.6 Installation

Note that following instructions are valid for Octave, version 4.0.3 only. GNU Octave can be downloaded [11] based on your operating system requirements. Various installation instructions are outlined at the wiki web site [12]. Installation is quite straightforward and user forums or a simple Google search yields useful answers to common problems encountered by users. From version number 3.8.1 onward, the Octave package installer comes with a default GUI interface. You should install this version for forthcoming discussions, although the older versions will prove to be fine as well.

1.6.1 Mac OSX

Two processes of installation are explained in this section. One uses a standard package installer provided by the Octave community for the Mac OSX operating system, which has a graphical instillation script. This must be run with sufficient privileges for proper installation. If you have issues, consult your system administrator. The other process uses `homebrew`, discussed shortly. It's good for those who love to install simply by using command line terminal. You should have sufficient knowledge of using a command line to use this method.

Installation Package

The installer can be downloaded [13] as a `.dmg` file. This file, when clicked, starts the GUI-based installation process where users can choose a location to install. A logo [14] representing Octave appears in the Application folder, and you simply click it to start the software. See Figure 1-1.

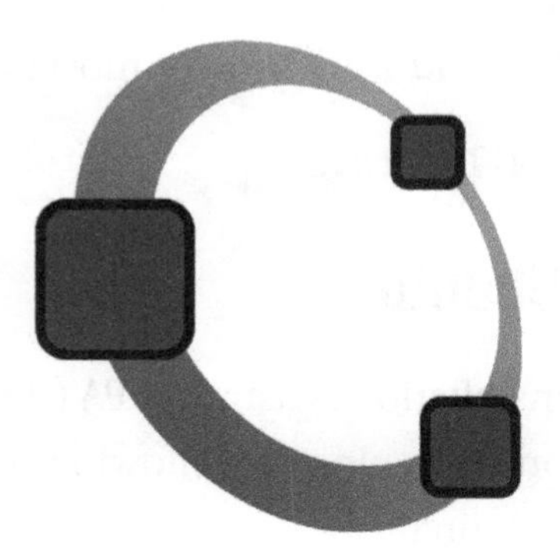

Figure 1-1. *The Octave logo*

Homebrew

An alternative is to *brew* the Octave software within Mac OSX using homebrew [15]. This involves first installing `homebrew` on your system and then simply issuing this command from the terminal:

```
1  $brew install octave
```

All its dependencies are installed and the latest version of the homebrew repository is properly installed. You must have administrative privileges for the account doing the installation. You can update to a newer version using this command:

```
1  $brew update octave
2  $brew upgrade octave
```

1.6.2 Octave on Ubuntu

Ubuntu is one of the most famous Linux distributions. Binary packages for Octave are provided by all versions of Debian and Ubuntu. These are very well tested binaries and should work best for most users.

```
1  $sudo apt-get install octave
```

Octave's PPA for Ubuntu

An alternative route to installation is using a PPA (Personal Package Archive). This can be done by typing the following commands successively at the Linux command terminal.

```
1  $sudo apt-add-repository ppa:octave/stable
2  $sudo apt-get update
3  $sudo apt-get install octave
```

After installation, you can simply run Octave by typing `octave` at the terminal. This will start the Octave terminal, which is `octave:1>`, as shown in Figure 1-2.

To exit an Octave session, you type `exit()` at the Octave command prompt.

If you want to use the GUI, you start Octave by typing `octave --force-gui` at the command prompt. There are a few other command-line options you can use while starting. They are given at reference [16], but they must be used by an experienced user who knows how to use Linux commands well.

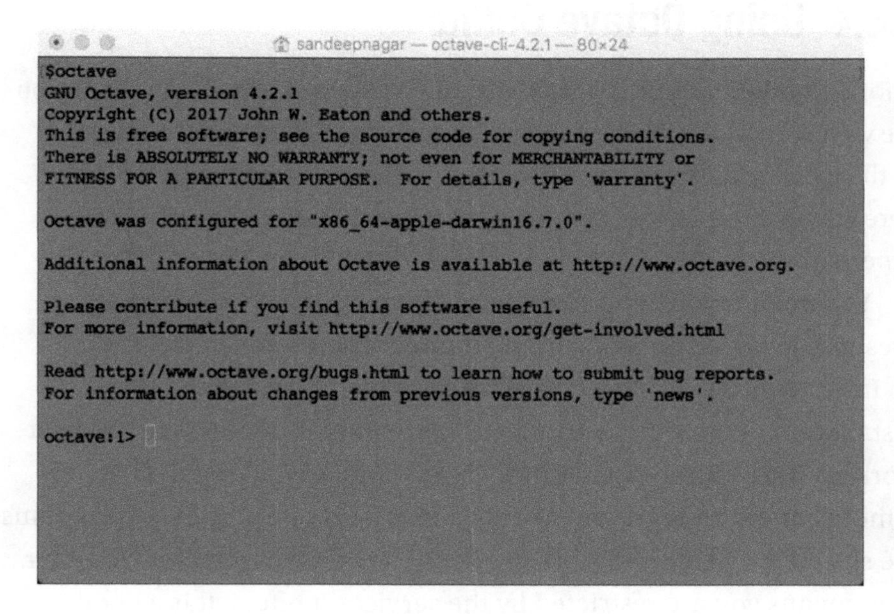

Figure 1-2. *Octave command-line interface*

1.6.3 Octave on Windows

Installation files for Windows OS can be found at the main web site [17]. The appearance of the Octave application is the same as shown in Figure 1-2. The color scheme for displaying Windows is based on the system preferences. It must be noted that users must first know whether their systems have a 32-bit or 64-bit architecture for the motherboard and hence the operating system. Accordingly, an installer must be chosen to avoid installation errors. Working within an Octave session will be uniform irrespective of the choice of operating system. But the OS-specific behaviors will be reflected in some places, like defining file paths. Windows- and Linux-based OSes use different separators for defining directory tree structure.

1.6.4 Using Octave Online

Another option for using Octave without an installation is to use it live on the web at [18]. See Figure 1-3. The basic form of Octave works as good as the local installation here. It is a good practice to start with Octave here and then proceeding with the local installation once you are more experienced.

You need to log in by creating an account first. Then you are presented with a window within an Internet browser. Since this is an Internet browser based installation, you don't need the local installation of the Octave application program. But note that the user works at a remote computer in this case. You upload the data at this remote computer and generate outputs here. Also, all Octave programs are stored in the limited storage offered by this free remote computer. That means you are restricted by the services it offers. It is a good practice for a beginner, but a serious developer will ultimately need a location installation too. This is especially the case when you do not have access to the Internet.

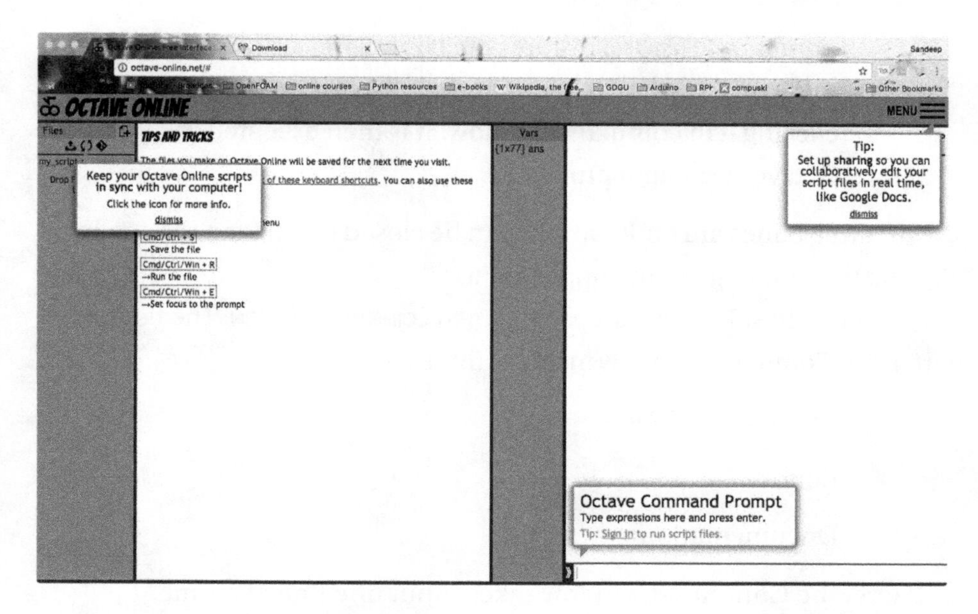

Figure 1-3. *Online instance of Octave at http://octave-online.net*

1.7 Octave GUI

Octave's GUI looks quite similar to MATLAB's GUI. As shown in Figure 1-4, the left side has three panels:

- *File Browser*: You can browse through the files in a working directory and change the names. You can run an `.m` file by clicking on the file. The file opens in the Editor window and can be run from there.
- *Workspace*: It stores the variables names, values, and their properties like types and sizes. It is useful for developers to visualize the variables and their contents. The meaning of variables and their values, sizes, etc. is illustrated in subsequent chapters.

- *Command History*: It stores the commands used in an Octave session. A command can be run by simply clicking it in command window. It is then executed at Octave command prompt.

All three panes are optional and can be closed down for a session by clicking the cross sign in the upper-right.

On the right side, there is a pane named `Command Window`. The bottom part of the Command window includes three tabs:

- Command window
- Editor
- Documentation

Here, the Command window takes input one line at a time. The Editor window is used to write an `.m` script file that can then be executed. This will be elaborated in Chapter 2 onward. The `Documentation` window can be used to read documentation and seek help to learn more about commands. Octave has an extensive documentation that enables a beginner to learn Octave with nothing but a command line. It also helps an experienced user who can seek help in using less common commands.

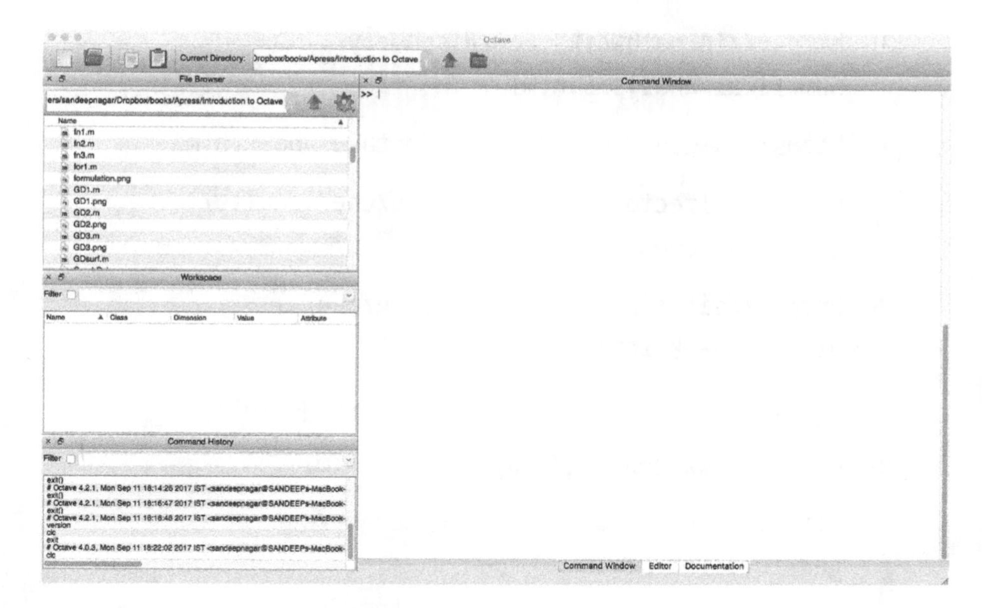

Figure 1-4. *Octave graphical user interface*

1.8 Summary

What started as a simple effort to provide students with a helpful set of programs to learn chemical reaction kinematics grew up as a major open source project that is now at par with its commercial counterpart. Octave evolved from human desire to share and advance collectively the knowledge base for the benefit of everyone. It further requires a consistent effort of budding developers working on this common aim. This book aims to provide a short and concise introduction to this software so that beginners can become users and ultimately developers of Octave.

1.9 Bibliography

[1] `https://mathworks.com/company/newsletters/articles/the-origins-of-matlab.html`

[2] `https://www.mathworks.com`

[3] https://in.mathworks.com/discovery/numerical-analysis.html

[4] https://www.gnu.org/software/octave/about.html

[5] https://directory.engr.wisc.edu/che/faculty/rawlings_james

[6] https://sites.utexas.edu/ekerdtgroup/dr-john-g-ekerdt/

[7] http://www.levenspiel.com/octave/resume.htm

[8] https://www.gnu.org/copyleft/gpl.html

[9] https://www.gnu.org/software/octave/get-involved.html

[10] https://octave.sourceforge.io/

[11] https://www.gnu.org/software/octave/download.html

[12] https://wiki.octave.org/Octave_for_macOS#Custom_Installation_Instructions

[13] https://sourceforge.net/projects/octave/files/Octave%20MacOSX%20Binary/2016-07-11-binary-octave-4.0.3/octave_gui_403_appleblas.dmg/download

[14] https://www.gnu.org/software/octave/img/octave-logo.svg

[15] https://brew.sh/

[16] https://www.gnu.org/software/octave/doc/v4.0.1/Command-Line-Options.html#Command-Line-Options

[17] https://www.gnu.org/software/octave/download.html

[18] https://octave-online.net/

CHAPTER 2

Interactive Octave Sessions

2.1 Introduction

When you start an Octave session, you can work in an interactive session in the sense that the Octave prompt >> waits for you to input a command, which will be executed as soon as you press the `Enter` key at the end of command. The Octave command prompt presents a full-featured interactive command-line commonly called REPL (read-eval-print loop).

The interactive shell of the Octave programming language is commonly called REPL because it:

- Reads what a user types
- Evaluates what it reads
- Prints out the return value after evaluation
- Loops back and does it all over again

This kind of interactive working environment proves especially useful for debugging. It also helps in prototyping a problem, where each step can be visualized for its output in a live fashion. You can check the results of the particular code as soon as you finish writing it. The way to work with

S. Nagar, *Introduction to Octave*, https://doi.org/10.1007/978-1-4842-3201-9_2

Octave's REPL is to write the code, analyze the results, and continue this process until the final result is computed. In addition to allowing quick and easy evaluation of Octave statements, REPL also showcases:

- *A searchable history*: You can press the Up and Down keys on the keyboard to browse through past commands instead of writing them again.
- *Tab completion*: You can simply press a Tab after writing a few letters for a command to auto-complete it. This avoids syntax errors. If more than one option matches when the Tab key is pressed, these options are displayed at the command prompt. This action is displayed here:

```
1   >> cl
2   cla        class     clear     clock       close
3   clabel     clc       clf       cloglog     closereq
4   >> clo
5   clock      cloglog   close     closereq
6   >> clock
7   ans =
8   2017.0000     9.0000       11.0000       20.0000
    48.0000       21.6951
9
```

When cl is printed and the Tab key is pressed, a series of commands that start with these letters are shown. When clo is typed and the Tab key is pressed, then the commands that start with clo are shown. Finally when clock is typed, an output is generated. The output prints the system time in year, month, date, time, minutes, and seconds.

- *Many helpful key bindings*: The key bindings depend on the operating system. When you click various items on the menu bar (at the top of Octave's main window), you will see the key bindings next to the name of the option.
- *Help and documentation*: Getting help on each topic and locating the documentation is also easy in Octave. You can simply feed any argument in as a string (i.e., characters enclosed within double quotes " ") to the built-in function `help()` or `doc()`.
  - Using `help()`, you'll get the help strings (useful illustrations of using a function/command) on the command line. An example is shown in Figure 2-1, where writing `help("version")` shows the usage of the `version` command.
  - Using `doc()`, you'll get detailed documentation under the documentation window instead of displaying it at the command prompt. The example shown in Figure 2-2 shows the usage of the `version` command.

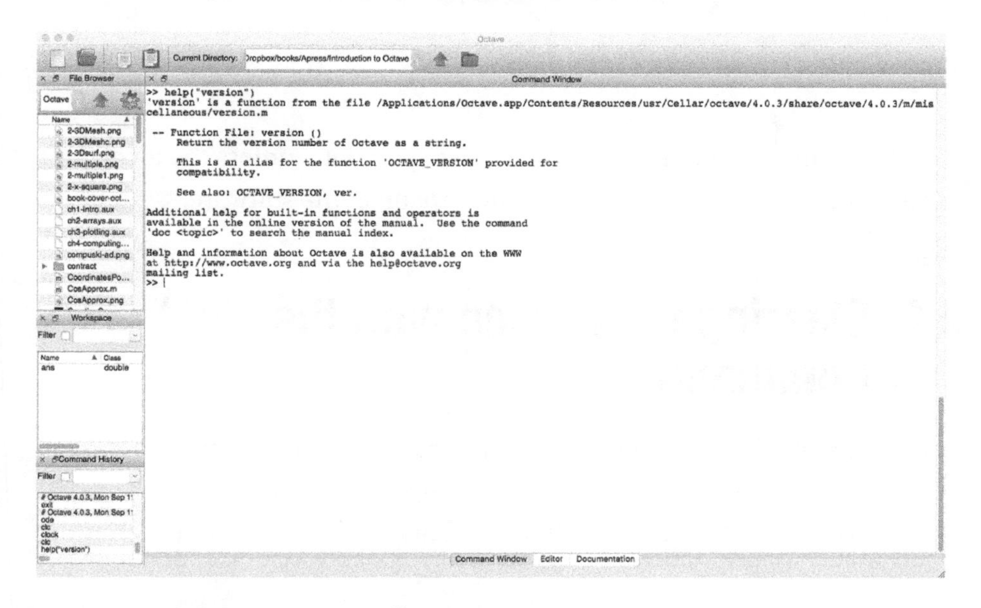

Figure 2-1. *Using the help() function*

Another way of seeking help is to use Octave's main menu bar, which has an option titled `Help`. You can type in the command there or use the submenu, titled `Documentation`. Documentation files can be found on your local disk as well as online.

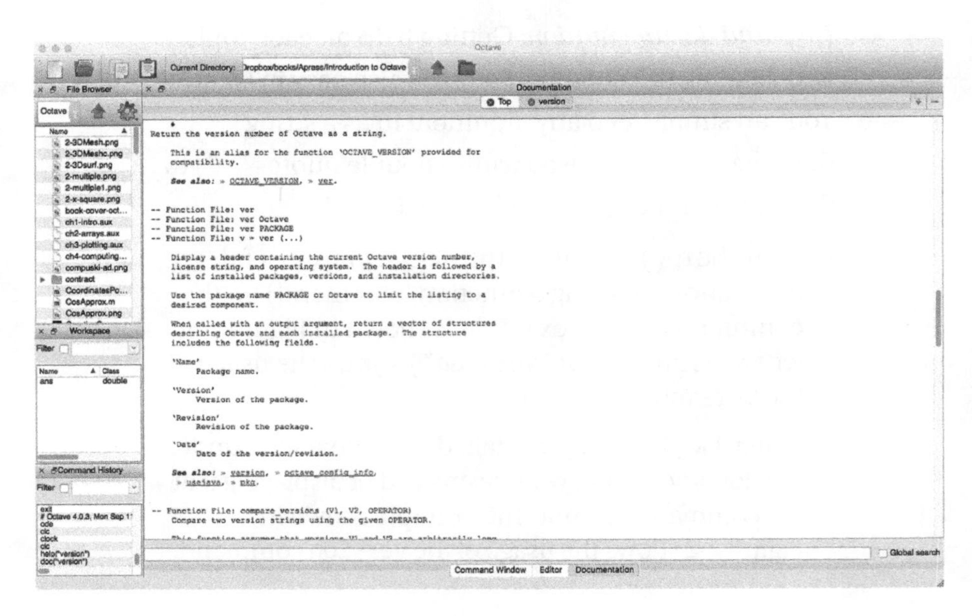

Figure 2-2. *Using the doc() function*

The online documentation is the most up to date, unless the local installation is using the most up to date version of the software.

2.2 Clearing the Screen with the clc( ) Command

Sometimes you'll need to obtain a clear screen, which is what the `clc` command does. You can learn about its proper usage by using the `help("clc")` or `doc("clc")` commands. This command presents a fresh command prompt. It is worth noting that Octave does not restart during this process. It merely shows a fresh screen to the user. It still stores all the variables and their respective values.

You can also use this command as a function by writing `clcl(n)`, where `n` is an integer. The command will clear that many number of lines from the previous session. If you simply type `clc()` instead, all lines are cleared from the screen.

2.3 Customizing the Octave Prompt

The `PS1()` function can be used to customize the Octave prompt to any desired string. For example, suppose you want to use `octave>>` as the Octave command prompt. You simply type the following command:

```
1  >> PS1("Octave>>")
2  Octave>> PS1(">>")
3  >>
```

Notice that the second command—`PS1(">>")`—is written at the changed command prompt—`PS1("Octave>>")`. Also, this returns the default command prompt. This book showcases the default prompt, but feel free to change it based on your choice and creativity.

If you want to be creative and have a unique command prompt, you can use one of the special characters [1] listed in Table 2-1.

Table 2-1. *Special Characters and Their Meanings in a String*

Special Characters	Meaning
\t	Time
\d	Date
\n	New line character
\h	Host name
\s	Name of program, i.e., octave
\w	Current working directory
\u	Username
#	Command number since Octave started
\\	A backslash

Using special characters, you can create a creative command prompt. Some examples are shown here:

```
1  >>PS1("\\u@\\h> ") % sets command prompt as username@
                        hostname
2  >>PS1("\\d@\\t> ") % sets command prompt as date@time
3  >>PS1("\\w@\\s> ") % sets command prompt as working
                        directory @program
```

Similarly, you can use a combination of options from Table 2-1 to create complex command prompts. You can always come back to the original command prompt by issuing the PS1(">>") command.

A second command prompt can be used by using the PS2() command. This happens when a user is prompted to input a value from the keyboard inside an Octave program. It is important to note that the particular command prompt is valid during a single session of Octave. When Octave restarts, the old values are lost and the default command prompt is issued at the Octave terminal.

Note that statements starting with `%` are treated as comments; they are not executed by Octave. Comments must be written to explain the structure of the code and beginners should practice writing comments for each line. They will help you understand your code after a long time and will also help others when the code is shared with them.

2.4 Working with Files

Apart from working on Octave REPL, you can write multi-line programs using the built-in text editor in Octave and run the program. Let's see how this works by writing a two-line program called `helloagain.m` (see Listing 2-1). This can be created by typing `edit helloagain` at the Octave command prompt. A new file called `helloagain.m` will be created in a folder/directory in which the present session of Octave is running. Alternatively, the program can also be created in the editor by clicking on the lowermost part of the `Command Window`, which has an option named `Editor`. This opens a blank editor window in which the `helloagain.m` code can be written manually. You can then save the file using `Ctrl+S`. By default. Note that all Octave script files are saved with an `.m` extension. You can open the existing file by navigating to the appropriate folder and choosing the file in the explorer.

Listing 2-1. The helloagain.m File

```
1  disp("\nHello World!\n")
2  disp("Hello again\n")
```

The \n character in the string input is used to print a newline character, which simply adds a paragraph return and prints the next characters on a new line. The disp() function prints the string at the command prompt.

You have many options for running an Octave file:

- You can simply type the name of file (without the extension) at the Octave command prompt. For example, the output of typing helloagain is shown here:

```
1  >> helloagain
2
3  Hello World!
4
5  Hello again
6
7  >>
8
```

- From the Editor menu, you can click on Run and choose Save File and Run. You can also choose to click the given key combination. It prompts you to save the file if the script file is being run for the first time. You can choose to save the file at a chosen destination within the local computer's storage.

In any case, the output is displayed at the command prompt, unless graphical output is directed to a graphical terminal. For now, we'll restrict the discussion to textual output.

These two methods of working with Octave (using REPL and using files) each has its own merits and usage. Interactive sessions are best for quickly checking for a small part of complex code. Files are best with a project involving detailed calculations and are linked with one another to perform a computational task. The following chapters discuss various concepts wrapped around these two kinds of sessions.

2.5 Using the Workspace

A *workspace* is the abstract space reserved for objects in the Octave session. All the objects used in calculations are displayed. This is usually placed as the second option in the left panel of the main Octave session window. You can observe that `ans` appears when anything is executed at the command prompt. For example, when `helloagain.m` was executed, `ans` was created and it stored `Char` (characters). If `ans` is now written at the command prompt, it prints the full path to the file `helloagain.m` in the local computer.

The `ans` displayed in the workspace is called a *variable* (because it can store a variety of values) and it stores the last executed values as a result of evaluating the expression at the Octave command prompt. All commands are treated as expressions at the Octave command prompt. The `ans` is called the global variable because it can be accessed globally, i.e., it has a global scope. This means that any Octave function can access it for usage and modification. More illustrations are shown in Chapter 3. The workspace window displays all global and *local* variables (those that have specific/local scope within a function only). The command `clear` clears all global and local variables in the workspace and makes it fresh, just as when an Octave session is initially launched. Its detailed usage is explained in Chapter 3.

2.6 Suppressing the Output Display

If the semicolon symbol `;` is used at the end of a command, the output is not displayed upon the execution of the command. The workspace is appropriately populated with local and global variables, their values, and other properties, but the output display is suppressed. This is useful when you expect too much output would be displayed. For example, when you are dealing with a multitude of data points, say a million data points, it would be pointless to invest time and computer memory in displaying them at Octave's command prompt. This feature can also be used within Octave scripts, when you don't want to print a particular output at the command prompt.

2.7 Running an Octave Program from the System Terminal

All operating systems offer terminal programs from which commands can be issued. A program can be run from a terminal by typing the name of application program stored in the system. In the case of Octave, this is `$octave`. Hence an Octave program, such as `helloagain.m`, can be executed from a system's terminal by issuing this command:

```
1  $octave helloagain.m
```

It is important to note that this works only if you are working in the same directory in which the program is stored. Otherwise, instead of just filename, you must provide the full file path. You can also run multiple Octave files by typing their names successively, separated by whitespace characters.

2.8 Summary

This chapter introduced working with Octave using single-line programs or multi-line script files. Working with Octave REPL in an efficient manner is a critical skill for a developer. At the same time, Octave also presents a feature rich built-in text editor. This makes life for an Octave developer quite easy, as you don't need to write the code outside the Octave session and then run it using an Octave program.

2.9 Bibliography

[1] https://www.gnu.org/software/octave/doc/v4.0.0/Customizing-the-Prompt.html

CHAPTER 3

Mathematical Expressions

3.1 Octave and Math

Octave is primarily a tool for solving mathematical problems numerically. This must necessarily mean that Octave provides a means to define mathematical structures in some way so that Octave REPL can *evaluate* them. Octave REPL must also understand some mathematical symbols like + (addition), - (subtraction), * (multiplication), and / (division) and their behavior must match their mathematical definitions. Octave must also be able to define a variety of numbers in its framework and be able to operate arithmetic operations on them appropriately. If Octave cannot depict a number system or cannot define mathematical operations, then you cannot perform those calculations using Octave. For example, Octave at present cannot perform `bra` and `ket` operator-based calculations for quantum mechanics.

Let's start probing Octave's abilities to perform simple arithmetic operations first and then dive into complex calculations. Keep in mind that the speed of execution depends on the user's machine's hardware. Older computers having less RAM and slower processors will take longer to execute a program than newer computers. Speed also depends on

S. Nagar, *Introduction to Octave*, https://doi.org/10.1007/978-1-4842-3201-9_3

the processor's availability for requests from Octave. Most of the newer computers perform calculations at lightening speeds, such that as soon as you press `Enter`, you can see the output of these simple calculations. But a lot of steps are happening behind the scenes and you must understand the process to optimally utilize both the computing resources at hand and Octave as a program.

3.2 Octave as a Calculator

In its simplest form, Octave works as a calculator with mathematical operators like multiplication (*), division (/), addition (+), subtraction (-), and exponentiation (^). The following code illustrates this behavior:

```
1  >> 3+5
2  ans =  8
3  >> 3.0+5.0
4  ans =  8
5  >> 3.1+5.0
6  ans =  8.1000
7  >> 2-3
8  ans =  -1
9  >> 3.0*5
10 ans =  15
11 >>  2/3
12 ans =  0.66667
13 >> format long
14 >> 2/3
15  ans =  0.666666666666667
16 >> format short
17 >> 2/3
18 ans =  0.66667
19 >> 2%3
```

```
20  ans =  2
21  >> 2^3
22  ans =  8
```

As you can see, when a command is entered at the Octave REPL command prompt `>>`, it is executed and an answer is displayed in the next line as `ans =`. As explained, `ans` is a global variable that stores the value of the last executed expression. The commands written at Octave REPL are called *expressions* and are evaluated by REPL. The behavior of this expression execution must be well defined to get a meaningful answer. Unlike other programming environments, even `3.0+5.0` yields an answer, which is displayed as an integer 8. But when `3.1+5.0` is evaluated, the answer is displayed as `8.1000`. To display more numbers in the result, you can use the `format long` command. By default, Octave works with the command `format short`.

In the section on data types later in this chapter, you learn that all objects belong to a certain data type. Integers and decimal numbers belong to two distinct data types and operators act accordingly. For example, adding two decimals is quite different than adding two integers. In the case of decimals, digits before and after the decimal point hold a different significance. They are also represented and stored differently in a computer.

Octave requires a * symbol to represent two or more multiplying entities. It is important to note that it is not `x` sign as you might expect. A computing system needs to know the operand and the operator. For example, * is an operator and `3.0` and `5` are the operators. Notice that if a result can be represented as an integer, Octave will do so. This behavior is quite different than other programming languages.

Dividing two numbers produces two outputs, the quotient and the remainder. Octave calculates the quotient using the `/` operator and calculates the remainder using the `%` operator. For this reason, when `2/3` is given to REPL, the quotient is evaluated as `0.66667`, and when `2%3` is given to REPL, the remainder is evaluated as 2.

By default, Octave displays five significant digits in human-readable form. The `format()` function can change this option. As seen in the following code, `format short` shortens the number of significant digits and `format long` lengthens them. But it is important to note that this does not alter the way Octave *stores* these numbers. They are stored as per their assigned/declared data type. The `format short e` and `format short E` commands both print the evaluated result in scientific notation.

```
1  >> format short e
2  >> 2/3
3  ans =   6.6667e-01
4  >> format short E
5  >> 2/3
6  ans =   6.6667E-01
```

This is inline with the fact that

$$\frac{2}{3} = 6.6667 \times 10^{-1} = 0.66667$$

Since the display is formatted for five significant digits, the last digit (i.e., the fifth one) is rounded toward $+\infty$. In both cases, the power of 10 is separated by either a lowercase `e` or an uppercase `E`. As discussed, the `short` command sets the significant digits to five. It can be set to 15 using `long`, as shown here:

```
1  >> format long e
2  >> 2/3
3  ans =   6.66666666666667e-01
4  >> format long E
5  >> 2/3
6  ans = 06.66666666666667E-01
```

If `short` g or `long` g is given to REPL, it chooses between a fixed point and exponential format based on the magnitude of the number. This behavior is shown here:

```
1  >> format short g
2  >> 2/3
3  ans =  0.66667
4  >> 2/3.1354222
5  ans =  0.63787
6  >> 2.12342/3.7773837383
7  ans =  0.56214
8  >> 29/9282829290200229
9  ans =  3.124e-15
10 >> format long g
11 >> 2/3
12 ans =  0.666666666666667
13 >> 2/3.1354222
14 ans =  0.637872628445381
15 >> 2.12342/3.7773837383
16 ans =  0.562140398517107
17 >> 29/9282829290200229
18 ans =  3.12404753910696e-15
```

If `short` eng or `long` eng is chosen, then the behavior is identical to `short` e or `long` e, except that the value is displayed using an engineering format, where the exponent is divisible by 3.

```
1  >> format short eng
2  >> 2/3
3  ans =  666.6667e-003
4  >> 29/9282829290200229
5  ans =  3.1240e-015
6  >> format long eng
```

```
7  >> 2/3
8  ans =   666.666666666663e-003
9  >> 29/928282929020229
10 ans =   3.12404753910696e-015
```

3.3 Rational Number Approximations

A rational number corresponds to a real number. But real numbers involve rational as well as irrational numbers. A real number can only be approximated as a rational expression. This can be achieved using the `rat()` function. An example is given as follows:

```
1 >> rat(2.34)
2 ans = 2 + 1/(3 + 1/(-17))
3 >> rat(2.3445643)
4 ans = 2 + 1/(3 + 1/(-10 + 1/(-4 + 1/(-2 + 1/(-2)))))
```

When these fractions are calculated and added, the final result is quite close to `2.34` and `2.3445643`, respectively. Since you may never achieve the exact number in most cases, this is called a *rational approximation.* The job of the `rat()` function to to print the rational approximation on the screen. Keep in mind that the rational approximation is merely printed this way; it is still stored as a real number (represented by floating point numbers). All the numerical outputs of REPL can be formatted as rational number approximations using the `format rat` command, as shown here:

```
1 >> format rat
2 >> 30.34
3 ans = 1517/50
4 >> 3.54/23.787986
5 ans = 4871/32732
```

The default is `format short`, which I suggest you use in the rest of the book. These formatting commands last only as long as the present Octave session. When Octave is restarted, the default settings are in place again.

3.3.1 Predefined Constants

A number of physical constants are defined as follows: `pi`, `e` (Euler's number), `i` and `j` (the imaginary number $\sqrt{-1}$), `inf` (infinity), and `NaN` (not a number, which results from undefined operations such as `Inf/Inf`).

```
1  >> pi
2  ans =  3.1416
3  >> e
4  ans =  2.7183
5  >> i
6  ans =  0 + 1i
7  >> j
8  ans =  0 + 1i
9  >> Inf/Inf
10 ans = NaN
```

They can also be used with formatted outputs, as shown in this code:

```
1  >> pi
2  ans =  3.1416
3  >> format long
4  >> pi
5  ans =  3.14159265358979
6  >> format long e
7  >> pi
8  ans =  3.14159265358979e+00
9  >> format short e
10 >> pi
```

```
11  ans =  3.1416e+00
12  >> format short g
13  >> pi
14  ans =  3.1416
15  >> format long g
16  >> pi
17  ans =  3.14159265358979
18  >> format short eng
19  >> pi
20  ans =  3.1416e+000
21  >> format long eng
22  >> pi
23  ans =  3.14159265358979e+000
24  >> format rat
25  >> pi
26  ans =  355/113
27  >> format short
```

It is interesting to note that π (which is an irrational number) can be depicted in a variety of formats on the terminal.

3.4 Using Complex Numbers

Computations involving complex numbers can be found in almost all branches of science and mathematics. The flexible way of defining complex numbers and their mathematics is an art that all Octave-based numerical computation developers must understand to compute efficiently. The world of complex numbers encompasses important scientific domains. When they are used to describe reality, they present more enriched pictures of physical phenomena as compared to using only real numbers.

Every programming language that can perform mathematical calculations robustly must handle complex numbers with ease. Octave is one such language. Complex numbers are defined with ease and most of their functions are present. Their usage in calculations with other data types is quite flexible and flawless.

3.4.1 Defining a Complex Number

A complex number can be defined in two ways:

- Using the `complex()` function with two inputs, where the first one is real and the second one is imaginary:

```
1  >> complex(2,3)
2  ans =  2 + 3i
3
```

- Straightaway as `a + ib` or `a + jb`:

```
1  >> i
2  ans =  0 + 1i
3  >> j
4  ans =  0 + 1i
5  >> 2 + 3i
6  ans =  2 + 3i
7  >> 2 + 3j
8  ans =  2 + 3i
9  >>
10
```

Both `i` and `j` hold the value of iota, i.e., $\sqrt{-1}$. The real and imaginary parts can be found using the `real()` and `imag()` functions, which take a complex number as their input.

```
1  >> real(complex(2,3))
2  ans =  2
3  >> imag(complex(2,3))
4  ans =  3
5  >> real(3)
6  ans =  3
7  >> imag(3)
8  ans =  0
```

When real and imaginary parts of a real number are probed, you obtain only the real part, as the imaginary part is zero.

3.4.2 Properties of Complex Numbers

Complex numbers are graphically defined as shown in Figure 3-1. On a real-imaginary axis based complex plane, a particular point is defined by a complex number $a + ib$, where a is magnitude of projection of the point on a real axis and b is magnitude of projection of the point on an imaginary axis.

The figure shows a point depicting the complex number $z = x + iy$. The values of $r = |z|$ (absolute value) and ϕ (argument) are given by:

$$r = \sqrt{x^2 + y^2} \quad \text{(Equation 3-1)}$$

$$\phi = tan^{-1}\left(\frac{y}{x}\right) \quad \text{(Equation 3-2)}$$

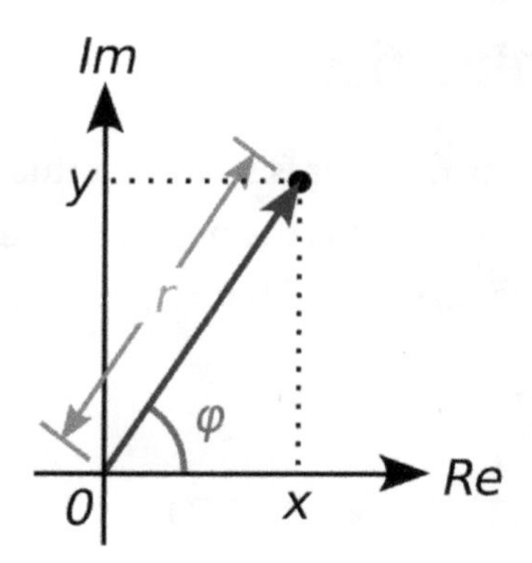

Figure 3-1. *Complex number depicted on complex plane [1]*

The absolute value of a complex number is simply its distance from the origin. The argument of a complex number is simply the angle it makes with the horizontal axis in a counterclockwise direction.

The principle and argument (in *radians*) for complex numbers—for example, $4 + 3i$ and $4 + 3i$—can be calculated using Octave.

```
1  >> abs(complex(4,3))
2  ans =  5
3  >> angle(complex(4,3))
4  ans =  0.64350
5  >> abs(complex(4,3))
6  ans =  5
7  >> angle(complex(4,3))
8  ans =  0.64350
9  >> abs(complex(-4,3))
10 ans =  5
11 >> angle(complex(-4,3))
12 ans =  2.4981
```

The angle outputs are produced in radians, which can then be converted into degrees. Here, we used the mathematical functions `abs()` and `angle()` on complex numbers. The next section illustrates more functions that can be used on real numbers too.

3.4.3 Using Conjugates

The *conjugate* of a complex number is its mirror image along the horizontal axis. Its imaginary part is the negative of the original number. When the original number is squared with its conjugate, you get r^2.

```
1  >> conj(complex(2,3))
2  ans =  2 - 3i
3  >> conj(complex(2,3))*complex(2,3)
4  ans =  13
```

3.4.4 Adding and Subtracting Two Complex Numbers

Complex arithmetic involves the typical operations such as addition, subtraction, multiplication, division, and exponentiation. However, the rules for complex numbers are a bit different.

Adding two complex numbers involves adding their real and imaginary parts. This is also the case with subtraction. Suppose you define two complex numbers as follows:

$$z_1 = a_1 + b_1 i$$

$$z_2 = a_2 + b_2 i$$

Then you can define their addition and subtraction as follows:

$$z_1 + z_2 = (a_1 + a_2) + (b_1 + b_2)i$$

$$z_1 - z_2 = (a_1 + a_2) - (b_1 + b_2)i$$

This can be verified by the Octave code as follows:

```
1  >> (2 + 3i)+(3 + 2i)
2  ans =  5 + 5i
3  >> (2 + 3i)-(3 + 2i)
4  ans =  -1 + 1i
```

3.4.5 Multiplying and Dividing Complex Numbers

Multiplication and division operations on complex numbers are not that straightforward. Consider this operation:

$$z_1 \times z_2 = (a_1 \times a_2) + (a_1 \times b_2)i + (a_2 \times b_2) + (b_1 \times b_2)(i^2)$$

It simplifies by collecting the real terms and imaginary terms, as follows:

$$z_1 \times z_2 = (a_1 a_2 - b_1 b_2) + (a_1 b_2 + a_2 b_1)i$$

because $i^2 = -1$. You can multiply and divide a complex number by a real number by simply performing the multiplication or division for the real and imaginary part, respectively.

A complex conjugate of a complex number, such as $z_1 = a_1 + b_1 i$, is defined as $z_1^* = a_1 - b_1 i$. Geometrically, z_1^* is the "reflection" of z_1 about the real axis. Hence, if you calculate the conjugate twice, you get the same number: $(z_1^*)^* = z_1$.

You can divide a complex number using its conjugate, as follows:

$$\frac{a_1 + b_1 i}{a_2 + b_2 i} = \frac{a_1 + b_1 i}{a_2 + b_2 i} \times \frac{a_2 - b_2 i}{a_2 - b_2 i} = \frac{a_1 a_2 + b_1 b_2}{a_2^2 + b_2^2} + \frac{b_1 a_2 - a_1 b_2}{a_2^2 + b_2^2} i$$

So multiplying the denominator's complex conjugate by both the numerator and the denominator yields a new complex number, which is the result of dividing two complex numbers.

```
1 >> (2 + 3i)*(2 - 3i)
2 ans =  13
3 >> (2 + 3i)*(4 - 4i)
4 ans =  20 + 4i
5 >> (2 + 3i)/(4 - 4i)
6 ans =  -0.12500 + 0.62500i
```

3.5 Common Mathematical Functions

A number of built-in mathematical functions exist in Octave. Describing each one is beyond the scope of this book, but a few of the more common ones are explained in this section.

- *Absolute value*: `abs()`.
- *Logarithm*: Natural logarithm `log()`, Base-10 logarithm `log10()`.
- *Trigonometric functions*: `sin()`, `cos()`, and `tan()`. Arguments are taken in radians.
- *Inverse-trigonometric functions*: `asin()`, `acos()`, and `atan()`.

```
1  >> abs(-10.034) // absolute value
2  ans =  10.034
3  >> log (e) // logarithm to base e
4  ans =  1
5  >> log10(10) // logarithm to base 10
6  ans =  1
7  >> sin(10) // sine(angle in radians)
8  ans =  -0.54402
9  >> cos(10) // cosine(angle in radians)
10 ans =  -0.83907
11 >> tan(10) // tangent(angle in radians)
12 ans =  0.64836
13 >> asin(1) // arcsin or inverse of sine
14 ans =  1.5708
15 >> asin(10)
16 ans =  1.5708 + 2.9932i
17 >> acos(1) // arccosin or inverse of cosine
```

```
18  ans =  0
19  >> acos(10)
20  ans =  0.00000 - 2.99322i
21  >> atan(1) // arctan or inverse of tangent
22  ans =  0.78540
23  >> atan(10)
24  ans =  1.4711
```

Complex calculations using these functions and operations can be performed with ease:

$$\sqrt{sin(10)^2 + cos(10)^2}$$

and

$$\frac{sin(10)}{\sqrt{cos(10)}}$$

```
1 >> sqrt(((sin(10))^2)+(cos(10))^2)
2 ans =  1
3 >> sin(10)/sqrt(cos(10))
4 ans =  0.00000 + 0.59390i
```

3.6 Learning More Mathematical Functions

Covering all the functions available in Octave is beyond the scope of this book. To understand how a particular function needs to be used, you can use the `help()` command, where the argument is the function you want to learn about. For example, `help(exp)` gives a detailed view of how this function should be used.

```
1  >> help("exp")
2  'exp' is a built-in function from the file libinterp/
   corefcn/mappers.cc
3
4  -- Mapping Function: exp(X)
5  Compute 'e^x' for each element of X.
6
7  To compute the matrix exponential, see *note Linear
   Algebra::.
8
9  See also: log.
10
11 Additional help for built-in functions and operators is
12 available in the online version of the manual. Use the command
13 'doc<topic>' to search the manual index.
14
15 Help and information about Octave is also available on the
16 WWW at http://www.octave.org and via the help@octave.org
17 mailing list.
```

Using the `doc("exp")` command, you will obtain detailed syntax information as well as example code in some cases. You can simply browse the *Arithmetic* section of the documentation to learn more.

3.7 Using Variables

Until now, we have been feeding numbers into Octave REPL with on-the-spot evaluation. Alternatively, you can designate a memory location where values are stored and this memory location can be known by a name for ease of usage. Such a programming construct is known as a *variable.*

To store values temporarily, you use variables that store the value at a particular memory location and address it with a symbol or set of symbols (called *strings*). For example, you can store the value of `1/10*pi` as a variable a and then use it in an equation like so:

$$a^2 + 10\sqrt{a}$$

```
1  >> a=1/10*pi
2  a =  0.31416
3  >> a^2 + 10*  sqrt(a)
4  ans =   5.7037
```

The symbol = works as an assignment operator because it assigns the value on the right side to the variable name on the left side. Its behavior is markedly different than its mathematical counterpart (which checks the equality of its right side and left side).

Multiple assignments can be performed using the comma (`,`) operator. Also if you do not want to produce results on-screen, you can suppress this by using the `;` operator.

```
1  >> a1 = 1, a2 = 10, a3 = 100
2  a1 =  1
3  a2 =  10
4  a3 =  100
5  >> a1 = 1, a2 = 10, a3 = 100;
6  a1 =  1
7  a2 =  10
8  >> a1 = 1; a2 = 10; a3 = 100;
9  >> a1
10 a1 =  1
11 >> a2
12 a2 =  10
13 >> a3
14 a3 =  100
```

3.7.1 Data Types

While assigning data to a variable, it is important to understand that data can be defined as a variety of objects defined by a data type, as follows:

- *Logical*: This type of data stores boolean values `1` or `0`. Boolean values can be operated on by boolean operators, like AND, OR, XOR, etc.
- *Char*: This type of data stores alphabetic characters and strings (groups of characters written in a sequence).
- *Int8, int16, int32, and int64*: This type of data is stored as integers within 8 bits, 10 bits, 32 bits, and 64 bits. The size of the integer is given by its bit counts. Both `logical` and `char` are 1 byte (8 bits) wide.
- *uint8, uint16, uint32, and uint64*: This type of data stores unsigned integer data in 8, 16, 32, and 64 bits.
- *double, single*: This type of data is stored as double and single precision floating types, respectively. Decimal numbers are represented by floating point data types. Single precision occupies 4 bytes (32 bits) and double precision occupies 8 bytes (64 bits) to store the floating point numbers.

  In a single precision system, 23 bits store the fraction bits (i.e., the numbers after the decimal point), 8 bits store the exponent (i.e., the numbers before the decimal point), and the 32nd bit is reserved for storing the sign.

  In a double precision system, 52 bits store the fraction bits (i.e., the numbers after the decimal point), 11 bits store the exponent (i.e., the numbers before the decimal point), and the 64th bit is reserved for storing the sign.

Single and double precision matters when the precision of the result matters. In cases like GPS position for a projectile flying at high speeds, the results should be as precise as possible for greater accuracy.

- *double complex, single complex*: Complex numbers have real and imaginary parts, which are stored separately. These numbers can be stored as single or double precision numbers using these data types.

3.7.2 Floating Point Numbers and Arithmetic

Real numbers are represented as floating-point numbers in a computer. The mapping of a real number to a computer's storage system is a formulaic representation (called a *floating point representation*) [2]. Here, real numbers are expressed in three parts—the significand, base, and exponent.

For example, the value of π is 3.1415926535897.... Let's suppose that you have only four significant digits for a particular calculation. So the value can be rewritten as 3.1415. Now this number is represented as $31415 \text{ x } 10^{-4}$ where 31415 is the significand, 10 is the base, and -4 is the exponent.

While assigning a number to the significand, the information about the number of significant digits is used. The significant figures of a number are the digits that carry meaning contributing to its measurement. In the current example, we assumed only four significant digits, depending on the requirements of the calculations/measurements. The term *floating point* refers to the fact that a number's *radix point* (decimal point) can float; that is, it can be placed anywhere relative to the significant digits of the number. This position is indicated as the exponent component, and thus the floating point representation can be thought of as a kind of scientific notation.

Storing a Floating Point Number

Computers can store numbers as floating point objects. A floating point object stores a number as follows:

$$\pm d_1 d_2 \cdots d_s \times \beta^e \tag{3.3}$$

Where $d_i = 0, 1, 2...\beta - 1$ but $d_1 \neq 0$ and $m \leq e \leq M$, where $m \in I^-$ and $M \in I^+$.

The three parts of a floating point number are:

- Sign ($\pm$)
- Significand (mantissa) ($d_1 d_2 ... d_s$)
- Exponent (β)

Each part of a floating point number is stored in different memory locations and occupies a specified number of bits. How many bits are defined to which parts? These questions have been answered by IEEE standards known as the IEEE754 [2]. First let's look at the concept of precision in a number representation.

- *Single precision*: Occupies 4 bytes/32 bits
- *Double precision*: Occupies 8 bytes/64 bits
- *Extended double precision*: Occupies 80 bits
- *Quadruple precision*: Occupies 16 bytes/128 bits

Each version has one bit reserved for depicting the sign of the number. Others bits are divided based on significand and exponent. Since all numbers are stored as binary numbers in a computer, the base is always 2. Depending on the number of bits available for storage, the maximum numeral value can be defined for a data type.

3.7.3 Overflow and Underflow Errors

If n bits are available to the significand, the maximum value can be 2^n. For the overall data type, if n bits are available for storage, and if one of them must be used for assigning the *sign bit*, then $2^n - 1$ is the maximum numeral value that can be stored by that data type. The limits are toward the both extremes (positive and negative numbers) for each data type.

Hence, crossing the limits results in *overflow* and *underflow* errors. When a number bigger than the biggest possible number is stored within a data type, it's called an *overflow error*. Similarly, when a smaller number is raised past its precision, *underflow errors* might occur. These types of errors can affect the final result drastically, especially when they propagate in a calculation.

It is important to perform back-of-the-envelop calculations for a particular problem to get an idea about the maximum and minimum numbers expected when running a program. Accordingly, you can assign data types based on your calculations. If you do not assign data types, Julia will assign them according to its own rules and this might incur precision errors as well as underflow and overflow errors.

3.7.4 Floating Point Numbers vs. Real Numbers

Keep in mind that floating point numbers are abstracts of real numbers. Sometimes this abstraction fails to represent the real numbers precisely.

It depends on the user's judgment whether this failure is insignificant. You can still confidently use floating point representation for calculations, keeping in mind the errors. A few examples will make this clearer:

If $a, b, n \in \Re \; \exists \; c = \frac{a+b}{n}$ such that $c \in \Re$, where $\Re$ represents a set of real numbers. This essentially says that between any two real numbers, there exists another real number.

But, this is not true for floating point numbers because floating point numbers are defined for a finite precision.

Because of this, floating point numbers are *approximations* of real numbers.

$\sqrt{7} \times \sqrt{7} - 7 = 0$, but Julia shows a finite small number for this calculation due to the finite precision nature of floating point numbers used to define $\sqrt{7}$:

```
1  >> format long
2  >> sqrt(7)*sqrt(7)
3  ans =   7.00000000000000
4  >> (sqrt(7)*sqrt(7))-7
5  ans =   8.88178419700125e-16
6
```

3.7.5 The eps() Function

The `eps()` (epsilon) function defines the smallest floating point number that can be defined for a machine. Hence, `t` is the machine precision for representing a floating point number on a particular machine. For example, on a 64-bit system, it comes out to be $2.22044604925031 \times 10^{-16}$. The `eps()` function takes a floating point number as input and outputs the next floating point number that can be represented by the machine.

```
1  >> eps
2  ans =   2.22044604925031e-16
3  >> eps(1)
4  ans =   2.22044604925031e-16
5  >> eps(10)
6  ans =   1.77635683940025e-15
```

Machine resolution generally decides the resolution of results and affects the accuracy of results. Storing numbers with better resolution is costly in terms of time (it will take more time to calculate a larger number of bits) and storage (the numbers themselves will occupy more bits).

3.7.6 Naming Conventions for Variables

Variable names have certain naming conventions that you must abide by to avoid errors.

- Names should not start with a number; however, numbers can be used anywhere afterwards.
- Variable names are case-sensitive.
- Keywords cannot be used as names.
- Names can include underscores (_).

While naming a variable, if you need to verify that the name is not a keyword, you can use the built-in function `iskeyword(name)`. Simply typing `iskeyword()` produces a list of keywords, as shown here:

```
1  >> iskeyword()
2  ans =
3  {
4  [1,1] = __FILE__
5  [2,1] = __LINE__
6  [3,1] = break
7  [4,1] = case
8  [5,1] = catch
9  [6,1] = classdef
10 [7,1] = continue
11 [8,1] = do
12 [9,1] = else
13 [10,1] = elseif
14 [11,1] = end
15 [12,1] = end_try_catch
16 [13,1] = end_unwind_protect
17 [14,1] = endclassdef
```

```
18  [15,1] = endenumeration
19  [16,1] = endevents
20  [17,1] = endfor
21  [18,1] = endfunction
22  [19,1] = endif
23  [20,1] = endmethods
24  [21,1] = endparfor
25  [22,1] = endproperties
26  [23,1] = endswitch
27  [24,1] = endwhile
28  [25,1] = enumeration
29  [26,1] = events
30  [27,1] = for
31  [28,1] = function
32  [29,1] = global
33  [30,1] = if
34  [31,1] = methods
35  [32,1] = otherwise
36  [33,1] = parfor
37  [34,1] = persistent
38  [35,1] = properties
39  [36,1] = return
40  [37,1] = static
41  [38,1] = switch
42  [39,1] = try
43  [40,1] = until
44  [41,1] = unwind_protect
45  [42,1] = unwind_protect_cleanup
46  [43,1] = while
```

3.7.7 List of Variables

A list of all variables can be obtained by using the commands who and whos. The who command simply presents the list of variables in the workspace, whereas whos presents the list with more detail, including the size of the variable, the number of bytes used to store the variable, and the variable type.

```
1  >> who
2  Variables in the current scope:
3
4  ans
5
6  >> whos
7  Variables in the current scope:
8
9  Attr Name        Size                     Bytes  Class
10 ==== ====        ====                     =====  =====
11      ans         42x1                       336  cell
12
13 Total is 42 elements using 336 bytes
```

By using who and whos, you can keep track of memory requirements. Judicious use of memory resources is especially important with Raspberry Pi based systems.

3.7.8 Global and Local Variables

A variable declared globally (within the main program) is known as a *global* variable, whereas a variable declared locally (within a function) is known as a *local* variable. When you define the global declaration statement, it remains the same regardless of any new definitions, unless you issue the clear command to clear the variable names and values from memory.

```
1  >> global a =1
2  >> global a = 2
3  >> a
4  a =  1
5  >> clear
6  >> who
7  >> whos
8  >> a=1
9  a =  1
10 >> a=2
11 a =  2
12 >>
```

As you can see, `a = 1` stays the same regardless of the next definition, `a = 2`. When the `clear` command is issued at the command prompt, all variable names and values are flushed from memory and the variable name can be used again. This time, if it is not defined as a `global` variable, then its value can be changed repeatedly. The `isglobal()`command lets you check if a variable name has been defined as a global variable.

Global variables are used to define constants during numerical calculations. Suppose you wanted all variables except a select few to change values. In this case, you would define those unchanging values as global variables by giving them the name of your choice. The predefined variables like `pi`, `e`, etc. have been defined in a similar manner.

3.7.9 The clear Function

As you saw in the previous section, the `clear` command flushes out variable names and their values from memory. But it proves to be much more useful than that. Whereas `clear all` is the same as `clear`, it can also be used to selectively wipe out variables and their values. Simply type `help clear` to get a detailed view of its use:

```
1  >> help clear
2  'clear' is a built-in function from the file libinterp/
   corefcn/variables.cc
3
4  -- Command: clear [options] pattern ...
5  Delete the names matching the given patterns from the
   symbol table.
6  The pattern may contain the following special characters:
7
8  '?'
9  Match any single character.
10
11 '*'
12 Match zero or more characters.
13
14 '[LIST]'
15 Match the list of characters specified by LIST. If the first
16 character is '!' or '^', match all characters except those
17 specified by LIST. For example, the pattern '[a-zA-Z]' will
18 match all lowercase and uppercase alphabetic characters.
19
20 For example, the command
21
22 clear foo b*r
23
24 clears the name 'foo' and all names that begin with the
   letter 'b'
25 and end with the letter 'r'.
26
27 If 'clear' is called without any arguments, all user-defined
```

```
28 variables (local and global) are cleared from the symbol
   table. If
29 'clear' is called with at least one argument, only the
   visible
30 names matching the arguments are cleared. For example,
   suppose you
31 have defined a function 'foo', and then hidden it by
   performing the
32 assignment 'foo = 2'. Executing the command 'clear foo'
   once will
33 clear the variable definition and restore the definition of
   'foo'
34 as a function. Executing 'clear foo' a second time will
35 clear the function definition.
36
37 The following options are available in both long and short
   form
38
39 '-all, -a' Clears all local and global user-defined
40 variables and all functions from the symbol table.
41
42 '-exclusive, -x'
43 Clears the variables that don't match the following
   pattern.
44
45 '-functions, - f'
46 Clears the function names and the built-in symbols names.
47
48 '-global, -g'
```

```
49 Clears the global symbol names.
50
51 '-variables, -v'
52 Clears the local variable names.
53
54 '-classes, -c'
55 Clears the class structure table and clears all objects.
56
57 '-regexp, -r'
58 The arguments are treated as regular expressions as any
59 variables that match will be cleared.
60
61 With the exception of 'exclusive', all long options can be used
62 without the dash as well.
63
64
65 Additional help for built-in functions and operators is
66 available in the online version of the manual. Use the
67 command 'doc<topic>' to search the manual index.
68
69 Help and information about Octave is also available on the
70 WWW at http://www.octave.org and via the help@octave.org
71 mailing list.
72 >>
```

Judicious use of the clear command proves to be a very powerful tool in managing memory requirements for a memory-intensive numerical calculation.

3.8 Summary

Octave provides an easy means of performing mathematical calculations. REPL is quite intuitive and easy to use, but you must consider the fact that you always get an *approximate* result. Knowing the machine precision and using data types judiciously will help you avoid underflow and overflow errors, which are critical to any numerical analysis task.

3.9 Bibliography

[1] https://en.wikipedia.org/wiki/File:Complex_number_illustration_modarg.svg
[2] http://grouper.ieee.org/groups/754/

CHAPTER 4

Working with Arrays

4.1 Introduction

Until now, we have considered storing only one value as a variable. However, there can be situations when a set of elements require similar processing. Then it would be wise to store them as an ordered set instead of creating separate variables for each data point. Octave defines an object named `Arrays` that can store an ordered set of elements. The elements of an array can have any data type, but most numerical computations deal with numerical elements, such as integers, floating point numbers, and complex numbers. A group of operations can be performed element-wise on an array by simply using the `a.(dot)` operator in front of a symbolic representation of the arithmetic operator. Other mathematical functions—like `sin()`, `cos()`, `asin()`, etc.—are already vectorized, which means they perform the operation on each element of the given array. This chapter deals with defining arrays and using them for scientific computation.

Arrays can be defined by simply enclosing elements in square brackets and separating them by comma operators or whitespace. For example:

```
1  >> a1 = [1,2,3]
2  a1 =
3  1   2   3
4  >> a2 = [1 2 3]
```


S. Nagar, *Introduction to Octave*, https://doi.org/10.1007/978-1-4842-3201-9_4

```
5  a2 =
6  1   2   3
7  >> a3 = [1;2;3]
8  a3 =
9  1
10 2
11 3
12 >> a4 = ["Sandeep";"Nagar";"Author"]
13 a4 =
14 Sandeep
15 Nagar
16 Author
17 >> a5 = ["Sandeep",1,2]
18 warning: implicit conversion from numeric to char
19 a5 = Sandeep
20 >> a5 = ["Sandeep","","Nagar"]
21 a5 = Sandeep Nagar
```

As seen in the example code, elements must be the same type. Numerical data types can be converted to each other, but characters and strings cannot be converted to numerical data. The first definition of array a5 produced a warning message and ignored the elements that do not match the first element's data type. If elements are strings, then the output is printed as a simple string at the terminal.

Note that the semicolon ; operator sends the element in the next row instead of the next column. This way, a 2D or 3D array can be created. The comma/whitespace operator will separate elements in the same row and ; will define the element in the next column.

This chapter deals with using arrays as matrices and performing mathematical calculations on an ordered set of numerical values. Manipulating matrices and using them to perform numerical calculations was the key feature of MATLAB that made it so popular. Octave also includes most of the options facilitated by MATLAB in this regard.

4.2 Arrays and Matrices

Matrices have become an integrated part of numerical computation when dealing with large quantities of data. For a 2D matrix, elements have a unique row and column index through which you can access them. Rows and columns can be attributed to different properties under study. This way, you can fit data for two properties as a matrix and then use these matrices for numerical calculations. For example, suppose an element of a row is defined as `1` if a compound is a conductor, `2` if it is a semiconductor, and `3` if it is an insulator. A row vector (a matrix composed of only one row) `[1 0 0 3 2 1 3 0 1 0 3 2 1]` has information about 13 compounds. In a numerical conductivity experiment involving the conductive nature of a compound, this row vector (a 13x1 matrix) can be utilized. A particular element of an array can be accessed using its index. For 1D arrays, it is simply the row/column number, whereas for 2D arrays, it's a combination of row number and column number. For example:

```
1  >> a = [1 0 0 3 2 1 3 0 1 0 3 2 1] % defined an array
2  a =
3  1   0   0   3   2   1   3   0   1   0   3   2   1
4  >> a(1)   % first element
5  ans =  1
6  >> a(2)   % second element
7  ans =  0
8  >> a(10)  % tenth element
9  ans =  0
10 >> a = [1 2 3;4 5 6] // defining a two dimensional array
11 a =
12
13 1   2   3
14 4   5   6
15
```

```
16 >> a(1,2) % element in row 1 and column 2
17 ans = 2
18 >> a(2,1) % element in row 2 and column 1
19 ans = 4
20 >> a(2,2) % element in row 2 and column 2
21 ans = 5
22 >> a(2,4) % element in row 2 and column 4
23 error: A(I,J): column index out of bounds; value 4 out of
       bound 3
```

When a 2D array is defined (separate elements in the same row by commas or whitespace and separate elements of two rows by ;), they can be accessed by giving an appropriate index with row and column indices separated by commas.

In this way, multi-dimensional matrices can be formed. The next sections illustrate the processes of creating even higher dimensional arrays, creating subarrays of lower dimensions, accessing elements of higher dimensional arrays, and slicing arrays to make another arrays from their elements. But first the following sections illustrate how a array having three elements in a row can be effectively used to represent a vector and three elements can be treated as coordinate points in a particular coordinate system.

4.3 Arrays as Vectors

We live in a 3D world and if we consider a Cartesian coordinate system then we need three numbers along the x, y, and z axes to point at a particular point. In this case, the point (0, 0, 0) represents the origin and it is fixed for a system. The point (1, 0, 0) represents moving one step forward in the x direction, while (0, 0, 1) represents moving one step forward in the z direction. In a similar fashion, (0, -1, 0) represents moving one step backward in the *y* direction. You can also move multiple steps at a time. For example, (2,-3,5) represents a point achieved by moving two steps forward in the *x* direction, three steps backward in the *y* direction, and five steps forward in the *z* direction.

Octave can define these coordinates as a row vector (an array having just row elements) or as a column vector (an array having just column elements). For example:

```
1  >> a1 = [1,0,0]
2  a1 =
3
4  1   0   0
5
6  >> a2 = [0,0,1]
7  a2 =
8
9  0   0   1
10
11  >> a3 = [0,-1,0]
12  a3 =
13
14  0  -1   0
15
16  >> a4 = [2,-3,5]
17  a4 =
18
19  2  -3   5
```

The variables `a1,a2,a3,a4,` and `a5` store arrays corresponding to Cartesian coordinates (1,0,0), (0,0,1), (0,-1,0), and (2,-3,5), respectively.

4.3.1 Coordinate Properties and Basic Transformations

Two coordinates can be added and subtracted to get a new coordinate. For example, (1,0,0) and (2,-1,2) can be added. The individual *x*, *y*, and *z* components are added to get a new coordinate (3,-1,2). This can be done in Octave as follows:

```
1  >> a1 = [1,0,0]
2  a1 =
3
4  1   0   0
5
6  >> a2 = [2,-1,2]
7  a2 =
8
9  2  -1   2
10
11 >> a1 + a2
12 ans =
13
14 3  -1   2
15
16 >> a2 + a1
17 ans =
18
19 3  -1   2
20
21 >> a1 - a2
22 ans =
23
24 -1   1  -2
```

```
25
26  >> a2 - a1
27  ans =
28
29  1  -1   2
```

It is important to note that while a1+a2 is the same as a2+a1, performing a1-a2 and a2-a1 does not yield the same results. This is in line with the rules of coordinate transformation.

Dividing a coordinate by a real number means taking a smaller step in that particular direction. For example, (1,1,1) can be divided by 2 to obtain (0.5,0.5,0.5), which moves half a step in each direction. Similarly, multiplying real numbers means taking steps that many number of times. For example, multiplying (1,1,1) by 2 would result in the coordinates (2,2,2). This moves two steps forward in each direction. Similarly, multiplying by a negative number means moving that many number of steps in a backward direction. This is shown in the following Octave code:

```
 1  >> a1 = [1,1,1]
 2  a1 =
 3
 4  1   1   1
 5
 6  >> 0.5*a1
 7  ans =
 8
 9  0.50000   0.50000   0.50000
10
11  >> a1/2
12  ans =
13
14  0.50000   0.50000   0.50000
```

```
15
16  >> 2*a1
17  ans =
18
19  2   2   2
20
21  >> -2*a1
22  ans =
23
24  -2  -2  -2
25
26  >> a1/-2
27  ans =
28
29  -0.50000    -0.50000    -0.50000
```

It is important to note that we have used row vectors, but column vectors can also be used for the same purpose. The convention is to use row vectors but it is not a rule.

4.4 Higher Dimensional Arrays/Matrices

Instead of just three elements, you can have any number of elements in an array. For example:

```
1  >> a = [1,2,3,4,5]
2  a =
3
4  1   2   3   4   5
5
6  >> a1 = [10,11,12,13,14]
```

```
7  a1 =
8
9  10   11   12   13   14
10 >> matrix22 = [1,2;3,4]
11 matrix22 =
12
13 1   2
14 3   4
15 >> matrix33 = [1,2,3;4,5,6;7,8,9]
16 matrix33 =
17
18 1   2   3
19 4   5   6
20 7   8   9
21 >> size(a)
22 ans =
23
24 1   5
25
26 >> size(matrix22)
27 ans =
28
29 2   2
30
31 >> size(matrix33)
32 ans =
33
34 3   3
```

As seen in the example code, an array can be understood as a matrix consisting of rows and columns. Thus, you can make a desired sized matrix. For example, `matrix22` is a 2*X*2 matrix and `matrix33` is a 3*X*3 matrix, whereas a is a 1X5 matrix. The first number when defining the size indicates the number of rows, whereas the second number indicates the number of columns. The comma (,) operator operates by defining the *next* element in the same row, whereas the semicolon (;) operator defines the numbers in the next line/row.

The size() function provides information about the elements in the rows and columns. In case of higher dimensional matrices, the size() function outputs the number of elements in each dimension.

If the number of elements in each row/column does not match, you'll get an error message:

```
1  >> right33 = [1,2,3;4,5,6;7,8,9]
2  right33 =
3
4  1   2   3
5  4   5   6
6  7   8   9
7
8  >>wrong33 = [2,3;4,5,6;7,8,9]
9  error: vertical dimensions mismatch (1x2 vs 1x3)
10 >> wrong33 = [1,2,3;4,5,6;8,9]
11 error: vertical dimensions mismatch (2x3 vs 1x2)
```

In array wrong33, the first row has only two elements; you can also say that column 3 has only two elements, which is why an error message showing a dimensional mismatch is displayed.

Elements of an array can be any data type, as defined in Chapter 3. All elements of an array can be set to a particular data type by using the commands as shown here:

```
1  >> x = uint32 ([1,65535])
2  x =
3
4  1  65535
```

```
5
6  >> x = uint64([1,65535])
7  x =
8
9  1  65535
10
11 >> x = int16([1,65535])
12 x =
13
14 1  32767
15
16 >> x = int32([1,65535])
17 x =
18
19 1  65535
20
21 >> x = int64([1,65535])
22 x =
23
24 1  65535
25
26 >> x = float([1,65535])
27 error:  'float' undefined near line 1 column 5
28 >> x = single([1,65535])
29 x =
30
31 1   65535
32
33 >> x = double([1,65535])
34 x =
35
```

```
36  1   65535
37
38  >> x = single([1.0,65535e10])
39  x =
40
41  1.0000e+00   6.5535e+14
42
43  >> x = double([1.0,65535e10])
44  x =
45
46  1.0000e+00   6.5535e+14
```

Line 14 shows that if the element is set to `int16`, it can store a maximum value of 32767, regardless of being commanded to store a value bigger than that. Hence, it becomes supremely important to understand the data type of the elements beforehand, in order to avoid errors in numerical calculations.

Keep in mind also that storing very small numbers in larger numbers of bits is a waste of memory. (Line 46 displays that the number 1 is stored as a double precision floating point number, which occupies 64 bits, whereby essentially 63 bits except the last one are all zeros.)

4.5 Operations on Arrays and Vectors

Operating on arrays involves two aspects:

- Operating on two or more arrays
- Element-wise operations

All arithmetic operators, such as +, -, *, /, %, ^, etc., can be used in both cases. When you need to do element-wise operations, then a dot `.` is placed before the operator. The element-wise operators therefore become `.+`, `.-`, `.*`, `./`, `.%`, and `.^`. This will become clearer in the following example.

```
1  >> a = [1,2;3,4]
2  a =
3
4  1    2
5  3    4
6
7  >> b = [5,6;7,8]
8  b =
9
10  5    6
11  7    8
12
13  >> a + b
14  ans =
15
16  6    8
17  10   12
18
19  >> 2.+ a
20  ans =
21
22  3    4
23  5    6
24
25  >> -10.+ b
26  ans =
27
28  -5   -4
29  -3   -2
```

When a and b are matrices to be added/subtracted, their elements are added/subtracted with elements in the same position. For this reason, the size of the two matrices added or subtracted must be the same.

We write `2.+a` and then add 2 to each element individually. This can be done regardless of size and is implemented uniformly on all the elements of the matrix.

4.5.1 Matrix Multiplication

Those who are familiar with matrix algebra know that matrix multiplication and division is not a straightforward task. An $a \times b$ matrix can only be multiplied by a $b \times c$ matrix, which results in an $a \times c$ matrix. This is performed by multiplying elements of rows with elements of columns to get new elements.

```
1  >> a = [1,2;3,4;5,6]
2  a =
3
4  1   2
5  3   4
6  5   6
7  >> a'
8  ans =
9
10 1   3   5
11 2   4   6
12
13 >> a*a'
14 ans =
15
16 5    11    17
17 11    25    39
```

```
18  17   39   61
19  >> a/b
20  ans =
21
22  0.050000   0.050000   0.050000
23  0.116667   0.116667   0.116667
24  0.183333   0.183333   0.183333
```

The command a' transposes the matrix a. That means that rows are made into columns and vice versa.

4.5.2 Matrix Division and Inverse of a Matrix

Performing division on a matrix involves *matrix inversion*. This can be achieved using the inv() function if the input matrix is a square matrix. Otherwise, the pin() function must be used. A square matrix has an equal number of rows in each dimension. The built-in function issquare() can be used to check if the given matrix (represented by an array) is a square matrix and whether an appropriate function should be used. The result is 1 if the matrix is a square matrix and 0 otherwise. Its usage is illustrated in the following code:

```
 1  >> a = [1,2;2,3]
 2  a =
 3
 4  1   2
 5  2   3
 6
 7  >> issquare(a)
 8  ans =  1
 9
10  >> b = [1,2;2,3;3,4]
```

```
11  b =
12
13  1   2
14  2   3
15  3   4
16
17  >> issquare(b)
18  ans = 0
```

When the inverse of a matrix is multiplied by itself, you get an identity matrix, i.e., a matrix with 1 as its elements in the diagonal direction and 0 everywhere else. This can be used to determine if the functions `inv()` and `pinv()` are working fine.

```
 1  >> a = [2,-2;4,2]
 2  a =
 3
 4  2  -2
 5  4   2
 6
 7  >> inv(a)
 8  ans =
 9
10  0.16667     0.16667
11  -0.33333    0.16667
12
13  >> a*inv(a)
14  ans =
15
16  1.00000     0.00000
17  0.00000     1.00000
18
```

```
19 >> b = [1,2;3,4;5,6]
20 b =
21
22 1   2
23 3   4
24 5   6
25
26 >> pinv(b)
27 ans =
28
29 -1.33333    -0.33333    0.66667
30 1.08333     0.33333     -0.41667
31
32 >> pinv(b)*b
33 ans =
34
35 1.0000e+00    8.8818e-16
36 -6.6613e-16   1.0000e+00
37
38 >> eps
39 ans =    2.2204e-16
```

pinv(b)*b results in very small numbers (of the order of 10^{-16}) instead of 0. This is of the same order of magnitude as the value of eps(). This means that these numbers can be approximated to be very close to zero, so the result is indeed an approximation of an identity matrix.

Identity Matrix

I is called an identity matrix because all of its diagonal elements are 1 and all its non-diagonal elements are 0, which makes its determinant 1.

The determinant of a matrix b is calculated by the command det(b). So let's investigate whether the determinant of pinv(b)*b is 1:

```
1  >> b = [1,2;3,4;5,6]
2  b =
3
4  1   2
5  3   4
6  5   6
7
8  >> pinv(b)
9  ans =
10
11  -1.33333     -0.33333     0.66667
12  1.08333      0.33333      -0.41667
13
14  >> pinv(b)*b
15  ans =
16
17  1.0000e+00    8.8818e-16
18  -6.6613e-16   1.0000e+00
19
20 >> det(pinv(b)*b)
21  ans =  1.0000
```

An identity matrix is automatically generated by using the command eye(a,b), where a and b are the number of rows and columns.

```
1  >> eye(2,2)
2  ans =
3
4  Diagonal Matrix
5
```

```
6 1 0
7 0 1
8 >> det(eye(2,2))
9 ans = 1
10 >> eye(4,5)
11 ans =
12
13 Diagonal Matrix
14
15 1 0 0 0 0
16 0 1 0 0 0
17 0 0 1 0 0
18 0 0 0 1 0
```

Division of Matrices

A matrix a can be divided by another matrix b by performing the following:

$$\frac{a}{b} = a \times b^{-1} \quad \text{(Equation 4-1)}$$

Let's test this behavior using Octave code, where first two matrices are created and stored in variables a and b. Then you can perform matrix division by calculating `a/b` and verify the same by calculating `a*pinv(b)`.

```
1 >> a = [1,2;3,4]
2 a =
3
4 1 2
5 3 4
6
7 >> b = [2,1;4,3]
8 b =
```

```
 9
10  2   1
11  4   3
12
13  >> a/b
14  ans =
15
16  -2.5000     1.5000
17  -3.5000     2.5000
18
19  >> a*inv(b)
20  ans =
21
22  -2.5000     1.5000
23  -3.5000     2.5000
```

These actions are used to determine a solution for a system of equations.

4.5.3 Finding Roots for a Set of Linear Equations

To practically apply these arrays to solve real-world problems, let's use arrays to find roots of a set of linear equations. Assume that you have the following:

$$x-4y-z=4 \quad \text{(Equation 4-2)}$$

$$-x-y+z=-1 \quad \text{(Equation 4-3)}$$

$$x+y-z=2 \quad \text{(Equation 4-4)}$$

You want to find those values of *x, y,* and *z* for which all three equations hold true. To do this, the first matrix formed from these equations can be written as:

$$\begin{bmatrix} 1 & -4 & -1 \\ -1 & -1 & 1 \\ 1 & 1 & -1 \end{bmatrix} \times \begin{bmatrix} x \\ y \\ x \end{bmatrix} = \begin{bmatrix} 4 \\ -1 \\ 2 \end{bmatrix} \quad \text{(Equation 4-5)}$$

If you assume that:

$$A = \begin{bmatrix} 1 & -4 & -1 \\ -1 & -1 & 1 \\ 1 & 1 & -1 \end{bmatrix} \quad \text{(Equation 4-6)}$$

$$X = \begin{bmatrix} x \\ y \\ x \end{bmatrix} \quad \text{(Equation 4-7)}$$

$$B = \begin{bmatrix} 4 \\ -1 \\ 2 \end{bmatrix} \quad \text{(Equation 4-8)}$$

Then you can write the following:

$$Ax = B \quad \text{(Equation 4-9)}$$

The solution is as follows:

$$X = A^{-1}B \quad \text{(Equation 4-10)}$$

This can be found with ease in Octave; you use just one command:

```
1  >> A = [1,-4,1;-1,-1,1;1,1,-1]
2  A =
3
4   1  -4   1
5  -1  -1   1
6   1   1  -1
7
8  >> B = [4;-1;2]
9  B =
10
11 4
12 -1
13 2
14
15 >> A\B
16 warning: matrix singular to machine precision
17 ans =
18
19 1.60526
20 -0.76316
21 -0.65789
```

Thus, $x = 1.60526$, $y = -0.76316$, and $z = -0.65789$, which satisfies the equations. In this way, Octave can perform complex matrix calculations with ease.

4.6 Summary

Array-based computing lies at the very heart of modern computational techniques. MATLAB became popular due to its ability to define computations in terms of matrix manipulations, which is reflected in its name (MATrix LABoratory). In a similar fashion, Octave presents a very suitable platform to perform this technique with ease. A variety of predefined functions enable users to save time when prototyping a problem. Flexible methods for defining multidimensional arrays and performing fast computations are critical these days. Most of the time spent on a simulation is in the loops or in array operations. Predefined array operations have been optimized with algorithms for reliability, time savings, and efficient memory management.

CHAPTER 5

Array Properties

5.1 Introduction

The preceding chapter discussed how arrays can effectively be used to represent a matrix and how matrix operations can be performed on array objects. This chapter discusses how arrays can be created automatically, how sub-arrays can be created from an array, and how arrays can be manipulated. These skills prove essential in the course of developing a numerical solution for a mathematical problem.

Automatic creation of arrays includes creating arrays of random numbers as well as creating arrays based on a formula. Pseudo-random numbers can be generated by algorithms defined in a base program. You can write your own algorithms too, but the base program presents optimized codes to perform such tasks. A number of types of random numbers can be generated based on the types of distributions. They should be chosen based on the requirements and this will be discussed.

Matrices can also be generated based on rules, like numbers from a starting point to a stopping point in linear fashion or logarithmic fashion. Matrices for zeros and one, as well as special matrices like upper triangular and lower triangular matrices, can be created with ease. The programs to create these matrices are written as Octave programs and supplied in the base package of Octave. You simply use them as regular Octave functions—you supply the inputs in a desired fashion and obtain the output. The `help()` and `doc()` functions can help you learn more about their use.

S. Nagar, *Introduction to Octave*, https://doi.org/10.1007/978-1-4842-3201-9_5

5.2 Automatic Creation of Arrays

Automatic creation of arrays is categorized into three areas:

- Creating random arrays, i.e., arrays with random numbers
- Creating arrays based on a rule
- Creating special matrices

When Octave is used to simulate real-world problems, sometimes you'll want to create fake arrays that are as close to reality as possible. This is where random number generators come in to play. It is worth noting that true random number generators do not exist, but most algorithms produce random numbers with very large cyclic repetitions and hence are good *pseudo*-random number generators.

5.3 Creating Random Matrices

Using random number generators, a random matrix can be created by the function `rand()`. The function `rand()` uses the Mersenne Twister with a period of 2^{19937} -1 [1]. It returns a matrix with random elements uniformly distributed on the interval (0, 1). Its usage is explained in the following code:

```
1  >> rand(1,5) // random matrix with 1 row and 5 columns
2  ans =
3
4  0.61623     0.23808     0.82978     0.99066     0.37742
5
6  >> rand(5,1) // random matrix of five rows and 1 column
7  ans =
8
```

```
 9  0.55830
10  0.51624
11  0.91662
12  0.74379
13  0.20169
14
15  >> rand(4,5) // random matrix of 4 rows and 5 columns
16  ans =
17
18  0.779821     0.904132     0.025018     0.118232     0.823903
19  0.963702     0.393643     0.148051     0.832420     0.316977
20  0.149530     0.943838     0.872814     0.699306     0.509816
21  0.133360     0.115337     0.401372     0.067246     0.264232
```

Note that the numbers generated here will be different each time, even on the same machine, since they are supposed to be random in nature. By default, they are uniformly distributed over the interval (0, 1). `help rand` gives a detailed description of the various other features and arguments of this random number generator.

By default, the generator is initialized from the `/dev/urandom` file if it is available; otherwise it's from the CPU time, wall clock time, and the current fraction of a second. This approach differs considerably from the approach used by MATLAB, which initializes to the same state at startup. The state of a random number generator is stored as a column vector of length 625. You can see this by typing `rand ("state")` on the terminal.

```
1  >>v = rand("state") % storing the state vector in variable v
2  >>rand("state", v) % setting the state vector to v
```

As mentioned, each time a random number is executed, you get a different number. This can become a problem if you want to generate the same set of random numbers. The state vector v (as stored in the previous code) can be used to generate the same set of random numbers. Setting the state to the one stored in v will always produce the same set of random numbers. The keyword reset can be used as an argument to reset the state vector of rand() so that it will again produce different sets of vectors. The following Octave code will make its usage clear.

```
1  >> v = rand("state"); % sets the state of rand() function to v
2  >> a1 = rand(3,4) % a1 stores 3 X 4 random number matrix
3  a1 =
4
5  0.56781   0.79619    0.85139   0.42739
6  0.72397   0.19870    0.96399   0.86126
7  0.31604   0.29627    0.54185   0.76511
8
9  >> rand("state",v); % sets the state to v
10 >> a2 = rand(3,4) % a2 stores 3 X 4 random number matrix
11 a2 =
12
13 0.56781   0.79619    0.85139   0.42739
14 0.72397   0.19870    0.96399   0.86126
15 0.31604   0.29627    0.54185   0.76511
16
17 >> a1 == a2 % a1 and a2 store similar numbers
18 ans =
19
20 1   1   1   1
21 1   1   1   1
22 1   1   1   1
23
```

```
24 >> rand("state", "reset"); % state is reset
25 >> a3 = rand(3,4) % a3 stores 3 X 4 random number matrix
26 a3 =
27
28 0.019441   0.141170     0.850737   0.145619
29 0.963360   0.888967     0.527707   0 558187
30 0.067530   0.228185     0.473682   0 625065
31
32 >> >> a3 == a2 % a3 elements are not similar to a2 since
                        the state has been reset
33 ans =
34
35 0   0   0   0
36 0   0   0   0
37 0   0   0   0
```

5.3.1 Creating Random Matrices with Integers

The function randi(imax) can be used in a similar fashion as rand() to produce integers from 1 to imax. The following Octave code will make the usage clear:

```
 1 >> randi(5)
 2 ans =  2
 3 >> randi(5)
 4 ans =  3
 5 >> randi(5)
 6 ans =  1
 7 >> randi(5)
 8 ans =  1
 9 >> randi(100)
10 ans =  81
```

```
11  >> randi(100)
12  ans =   60
13  >> randi(100)
14  ans =   21
15  >> randi(100)
16  ans =   31
17  >> randi(100)
18  ans =   17
```

randi(5) generates a random integer between 1 to 5 each time it is used. Similarly, rand(100) generates a random integer between 1 to 100.

A matrix of numbers can be constructed in a similar fashion, where the following arguments to the function define the number of elements in each dimension.

```
 1  >> randi(100,2,3) % matrix of 2 rows and 3 columns with
                        random integer from 1 to 100
 2  ans =
 3
 4  28    16    26
 5  36    90    79
 6
 7  >> randi(100,2,3) % generates different set of integers
 8  ans =
 9
10  41    32    45
11  3    49    91
```

The lower and upper bounds can also be defined as the first argument. This must be done as an array of two numbers, where the first element defines the lower bound and the second one defines the upper bound.

```
1  >> randi([-100,100],2,3) % 2 X 3 matrix of numbers from
                              -100 to +100
2  ans =
3
4  78   11   -59
5  -28  -74   -15
6
7  >> randi([-100,100],2,3) % A different set of numbers is
                              generated
8  ans =
9
10 29   45  -16
11 -20   75   34
```

The same set of random numbers can also be generated in a similar fashion as that of the rand() function. In fact, the state is set using the same function.

```
1  >> v1 = rand("state"); % storing state vector
2  >> randi([-100,100],2,3) % Generating 2 X 3 vector
3  ans =
4
5  14   71  -18
6  -27   49   18
7
8  >> rand("state",v1); % restoring the same state vector
9  >> randi([-100,100],2,3) % same set of elements are generated
10 ans =
11
12 14   71  -18
13 -27   49   18
```

5.3.2 Defining Random Numbers from a Set Distribution

Now you need to learn how to define random numbers based on different kinds of distributions. For this purpose, the following functions are used:

- `rande()`: Exponential distribution
- `randn()`: Normal distribution
- `randp()`: Poisson distribution
- `randg()`: Gamma (A,1) distribution

Choose your distributions judiciously. Choosing the right distribution for your simulation will make it more realistic.

The rande() Function

The `rande()` function returns a matrix with exponentially distributed random elements. The arguments are handled the same way as the arguments for `rand()`.

```
1  >> rande(4,5)
2  ans =
3
4  0.526399     0.586847     2.761980     1.006396     1.909515
5  1.496118     0.976633     0.059666     3.201508     0.898904
6  1.559492     0.266075     0.346443     0.129497     1.556362
7  0.281763     2.006331     0.892212     0.650638     0.651668
```

The randg() Function

This function returns a matrix with "gamma (A,1)" distributed random elements. The gamma distribution is a two-parameter family of continuous probability distributions. The common exponential distribution and chi-squared distribution are special cases of the gamma distribution.

```
1  >> randg(4,5)
2  ans =
3
4  5.25552     3.20796     9.40051     4.53603     7.30682
5  0.92867     5.48899     3.09422     4.26167     10.41489
6  2.35244     2.34990     1.26921     4.87652     3.97179
7  3.53832     3.12559     5.08525     6.50268     5.69221
8  4.98024     3.51875     9.43768     8.82720     5.74206
```

The randn() Function

This function returns a matrix with normally distributed random elements having a mean equal to 0 and a variance equal to 1. `randn()` uses the Marsaglia and Tsang Ziggurat technique [2] to change from a uniform to a normal distribution.

```
1  >> randn(4,5)
2  ans =
3
4  0.695713     2.013552    -0.076682    -0.695119    -0.889084
5  -1.659300    0.875251     0.385765     0.596478    -1.302996
6  -0.330802    1.554179     0.174712    -1.087671    -1.371431
7  -1.446307   -0.969824    -0.123708     1.014428     0.673549
```

The randp() Function

This function returns a matrix with Poisson distributed random elements with a mean value parameter given by the first argument. For example, if first argument is 1, then random numbers within the Poisson distribution having a mean of 1 are produced.

```
1  >> randp(1)
2  ans =  2
3  >> randp(1)
4  ans = 0
5  >> randp(1)
6  ans = 0
7  >> randp(1)
8  ans =  2
9  >> randp(1)
10 ans = 0
11 >> randp(1)
12 ans =  1
13 >> randp(1)
14 ans =  3
```

On the other hand, a matrix can also be produced by giving dimensions of the matrix as other arguments. Arguments can be presented as numbers separated by commas or as an array having a description of dimensions.

```
1 >> randp(1,2,3) % mean=1, matrix of 2 rows and 3 columns
2 ans =
3
4 1   0   1
5 0   0   1
6
```

```
7  >> randp(1,2,3) % repeating the same command and getting a
                      different set of numbers
8  ans =
9
10 1   3   1
11 1   1   2
12
13 >> randp(2,2,3) % mean=2, matrix of 2 rows and 3 columns
14 ans =
15
16 1   3   2
17 2   0   1
18
19 >> randp(2,2,3) % repeating the same command and getting a
                      different set of numbers
20 ans =
21
22 3   0   0
23 2   3   0
24
25 >> randp(20,4,3) % mean=20, matrix of 4 rows and 3 columns
26 ans =
27
28 21   17   22
29 23   21   13
30 17   24   12
31 30   13   13
32
33 >> randp(20,4,3) % repeating the same commands and getting
                       different set of numbers
34 ans =
35
```

```
36  25   29   27
37  28   11   20
38  17   22   36
39  21   16   18
40
41  >> randp(20,[4,3]) % Inputting matrix dimensions as an
                          array (4 rows and 3 columns)
42  ans =
43
44  18   18   19
45  29   22   13
46  19   25   24
47  20   18   17
48
49  >> randp(20,[3,4]) %% Inputting matrix dimensions as an
                           array (3 rows and 4 columns)
50  ans =
51
52  19   19   25   28
53  21   17   10   29
54  18   20   19   24
```

5.4 Automatic Generation of Large Arrays

You can automatically generate an array by defining a rule using the colon : operator or by using the `linspace()` and `logspace()` arguments. These methods are widely used, as they are convenient ways to generate large matrices. It is important to remember that you can suppress the output being printed on the terminal by ending the command with the semicolon ; operator, since it can be quite annoying to see a large set of numbers on the terminal.

5.4.1 Generating Arrays Using a Rule

You can generate a series of numbers and store them as arrays by using the command `start:step:stop`, where the numbers representing `start`, `step`, and `stop` are real numbers. When complex numbers are entered, they are converted to real numbers (only the real part of a complex number is used). The result is an array. Defining the brackets ([ ]) is optional. If the step is not defined, then it is taken as 1.

```
1  >>x = 1:10 % start 1 and stop 10, default step=1 is used
2  a =
3
4  1    2    3    4    5    6    7    8    9    10
5
6  >> x = 1:2:10 % without brackets, start=1, step=2, stop=10
7  x =
8
9  1    3    5    7    9
10
11 >> x = [1:1:10] % with brackets, start=1, step=1, stop =10
12 a =
13
14 1    2    3    4    5    6    7    8    9    10
15
16 >> y = 2.2:3.8 % start=2.2, stop=3.8, step=1
17 y =
18
19 2.2000    3.2000
20
21 >> y = 2.2:0.2:3.8 % start=2.2, step=0.2, stop=3.8
22 y =
23
```

```
24  2.2000    2.4000    2.6000    2.8000    3.0000
    3.2000    3.4000    3.6000    3.8000
25
26  >> a = 2 + 3i % Defining a complex number
27  a =  2 + 3i
28
29  >> a:a/2:a*2 % start=complex number a, step=a/2, stop=2a
30  ans =
31
32  2   3   4
33
34  >> a:a/3:a*3 % start=complex number a, step=a/3, stop=3a
35  ans =
36
37  2.0000    2.6667    3.3333    4.0000    4.6667
    5.3333    6.0000
```

5.4.2 Creating Linearly Spaced Vectors

The `linspace(start,stop,n)` command produces an array starting with the first number and stopping at the second one, with a total of `n` numbers. Hence, they are linearly spaced. When complex numbers are used as arguments, you get an array of complex numbers as the output. In this case, the step is calculated to be:

$$s = \frac{S+E}{n} \qquad \text{(Equation 5-1)}$$

Where s = step, S = start, E = End, and n = number of items.

```
1  >> a = linspace(1,2,5) % start=1, stop=2, number of items=5
2  a =
3
4  1.0000    1.2500    1.5000    1.7500    2.0000
5
6  >> a = linspace(1,2,10) % start=1, stop=2, number of items=10
7  a =
8
9 1.0000    1.1111    1.2222    1.3333    1.4444    1.5556
  1.6667    1.7778    1.8889    2.0000
10
11  >> a = 2 + 3i % defining the complex number
12  a =  2 + 3i
13  >> linspace(a,2*a,10) % start=a, stop=2a, number of items=10
14  ans =
15
16  Columns 1 through 6:
17
18  2.0000 + 3.0000i    2.2222 + 3.3333i    2.4444 + 3.6667i
    2.6667 + 4.0000i    2.8889 + 4.3333i    3.1111 + 4.6667i
19
20  Columns 7 through 10:
21
22  3.3333 + 5.0000i    3.5556 + 5.3333i    3.7778 + 5.6667i
    4.0000 + 6.0000i
```

5.4.3 Creating Logarithmically Spaced Vectors

Similar to `linspace`, the `logspace(A,B,n)` function returns a row vector with n elements, which are logarithmically spaced from 10^A to 10^B. This is useful in generating fictitious data involving exponential functions because data points must increase exponentially. When complex numbers are used as an input to this function, you obtain an array of complex numbers.

```
1  >> logspace(1,10,5)
2  ans =
3
4  1.0000e+01  1.7783e+03  3.1623e+05  5.6234e+07  1.0000e+10
5
6  >> logspace(1,-10,10)
7  ans =
8
9  1.0000e+01    5.9948e-01    3.5938e-02    2.1544e-03
   1.2915e-04    7.7426e-06    4.6416e-07    2.7826e-08
   1.6681e-09    1.0000e-10
10
11 >> logspace(1,10,10)
12 ans =
13
14 1.0000e+01     1.0000e+02     1.0000e+03     1.0000e+04
   1.0000e+05     1.0000e+06     1.0000e+07     1.0000e+08
   1.0000e+09     1.0000e+10
15
16 >> a = 2 + 3i
17 a =  2 + 3i
18 >> logspace(a,2*a,10)
19 ans =
20
```

```
21  Columns 1 through 5:
22
23  81.121 + 58.475i   29.650 + 164.154i   -154.540 + 231.395i
    -453.528 + 98.772i  -658.825 - 406.736i
24
25  Columns 6 through 10:
26
27 -319.756 - 1251.342i      1065.540 - 1872.489i
   3447.917 - 1013.589i      5312.927 + 2776.859i
   3161.384 + 9487.131i
```

5.5 Creating Special Matrices

Matrix algebra defines matrices that are special in nature and find their use in some specialized problems. Octave has some functions defined to create these matrices.

5.5.1 Upper and Lower Triangular Matrix

An upper triangular matrix is one where only the diagonal and the elements above diagonal are non-zero. Similarly, a lower triangular matrix is one where the diagonal and the elements below diagonal are non-zero. The `tril()` function returns a lower triangular matrix and `triu` returns an upper triangular matrix. They take input from another matrix and return a matrix of similar dimensions, but with modifications.

```
1  >> a = rand(3,3)
2  a =
3
4  0.414936     0.399589     0.269880
5  0.070691     0.405602     0.378955
```

```
6  0.169398     0.850042     0.919782
7
8  >> tril(a)
9  ans =
10
11 0.41494   0.00000   0.00000
12 0.07069   0.40560   0.00000
13 0.16940   0.85004   0.91978
14
15 >> triu(a)
16 ans =
17
18 0.41494   0.39959   0.26988
19 0.00000   0.40560   0.37896
20 0.00000   0.00000   0.91978
```

5.5.2 Diagonal Matrix

Using the diag() function, you can return an array of diagonal elements. The first argument is the matrix and the second argument indicates the direction of movement from the central diagonal, which is represented by the default value 0.

```
1  >> a = rand(4,4)
2
3  >> a
4  a =
5
6  0.159507     0.612608     0.962059     0.774479
7  0.571956     0.302159     0.933308     0.621334
8  0.024959     0.643726     0.043745     0.171901
9  0.136112     0.943376     0.056256     0.102074
10
```

```
11 >> diag(a) % central diagonal element
12 ans =
13
14 0.159507
15 0.302159
16 0.043745
17 0.102074
18
19 >> diag(a,1) % diagonal element after moving one step
                 upwards from central diagonal
20 ans =
21
22 0.61261
23 0.93331
24 0.17190
25
26 >> diag(a,-1) % diagonal element after moving one step
                  downwards from central diagonal
27 ans =
28
29 0.571956
30 0.643726
31 0.05625632
32
33 >> diag(a,-2) % diagonal element after moving two steps
                  downwards from central diagonal
34 ans =
35
36 0.024959
37 0.943376
38
```

```
39 >> diag(a,2) % diagonal element after moving two step
                 upwards from central diagonal
40 ans =
41
42 0.96206
43 0.62133
```

5.5.3 Ones and Zeros Matrices

A matrix that contains all 1s or all 0s is a ones matrix and zeros matrix, respectively:

```
 1 >> ones(3,3)
 2 ans =
 3
 4 1   1   1
 5 1   1   1
 6 1   1   1
 7
 8 >> zeros(3,3)
 9 ans =
10
11 0   0   0
12 0   0   0
13 0   0   0
```

These are generally used for initialization of matrices of desired dimensions. The initialized matrix is then used for manipulations.

5.5.4 Sparse Matrix

Since arrays can store large amounts of data, they can become so large that they are an issue for computers with limited storage capabilities. In some cases, many of these values are 0. In this case, it makes sense to have a special matrix to handle this class of problems, whereby only the non-zero elements of the matrix are stored.

This provides two aspects of an efficient computational framework. First, it reduces the amount of memory needed to store a matrix. It also means that you can take advantage of prior knowledge about positions of non-zero elements and devise the mathematical operations for targeted indices.

One of the simplest ways to store a sparse matrix is by storing the elements of the matrix as triplets:

- Two elements being their position in the array (i.e., their row and column indices).
- Third element being the data itself.

This is conceptually easy to grasp, but requires more storage than is strictly needed. Octave instead uses a compressed column format. The position of each element in a row and the data are stored the same way as the previous method, but the number of non-zero elements in each column is stored rather than their positions. This reduces the storage memory requirements.

You can create a sparse matrix in many ways. The `speye()` function returns an sparse identity matrix, whereby inputs define the number of elements in a dimension.

```
1  >> a = speye(2,3) % 2 X 3 sparse matrix
2  a =
3
```

```
4  Compressed Column Sparse (rows = 2, cols = 3, nnz = 2 [33%])
5
6  (1,1) -> 1
7  (2,2) -> 1
8
9  >> a = speye(20,30) % 20 X 30 sparse matrix
10 a =
11
12 Compressed Column Sparse (rows = 20, cols = 30, nnz = 20 [3.3%])
13
14 (1,1) -> 1
15 (2,2) -> 1
16 (3,3) -> 1
17 (4,4) -> 1
18 (5,5) -> 1
19 (6,6) -> 1
20 (7,7) -> 1
21 (8,8) -> 1
22 (9,9) -> 1
23 (10,10) -> 1
24 (11,11) -> 1
25 (12,12) -> 1
26 (13,13) -> 1
27 (14,14) -> 1
28 (15,15) -> 1
29 (16,16) -> 1
30 (17,17) -> 1
31 (18,18) -> 1
32 (19,19) -> 1
33 (20, 20) -> 1
```

The mathematical output of speye() and eye() is the same, but the amount of storage required is quite large in the case of the eye() function, because it stores all zeros in non-diagonal positions too.

The spdiags() function is a generalization of the diag() function, whereby the diagonals are non-zero. The first argument is the input matrix and the second argument is the direction of movement from the central diagonal.

```
1  >> a = rand(3,3)
2  a =
3
4  0.73015    0.35654    0.68810
5  0.43991    0.71976    0.51030
6  0.86486    0.95695    0.20814
7
8  >> spdiags(a)
9  ans =
10
11 0.86486    0.43991    0.73015    0.00000    0.00000
12 0.00000    0.95695    0.71976    0.35654    0.00000
13 0.00000    0.00000    0.20814    0.51030    0.68810
14
15 >> spdiags(a,2)
16 ans =
17
18 0.00000
19 0.00000
20 0.68810
21
22 >> spdiags(a,1)
23 ans =
24
```

```
25  0.00000
26  0.35654
27  0.51030
```

The sprand() function can be used to produce a sparse matrix of *uniformly* distributed random numbers. The first two arguments are the number of rows and columns, while the third argument determines the density (ith should be between 0 and 1). In a similar fashion, sprandn() creates *normally* distributed random numbers.

```
1  >> sprand(2,3,0.5) %2X3 matrix with 50% density
2  ans =
3
4  Compressed Column Sparse (rows = 2, cols = 3, nnz = 3 [50%])
5
6  (1,2) ->    0.45900
7  (2,2) ->    0.48153
8  (2,3) ->    0.58828
9
10 >> sprand(2,3,0.1) % 2X3 matrix with 10% density
11 ans =
12
13 Compressed Column Sparse (rows = 2, cols = 3, nnz = 1 [17%])
14
15 (2, 1) ->    0.37447
16
17 >> sprand (2,3,0.9) % 2X3 matrix with 90% density
18 ans =
19
20 Compressed Column Sparse (rows = 2, cols = 3, nnz = 5 [83%])
21
```

```
22 (1,1) ->   0.57181
23 (1,2) ->   0.31887
24 (2,2) ->   0.45919
25 (1,3) ->   0.69767
26 (2,3) ->   0.76126
```

Note from this Octave code that density desired and density obtained do not match exactly. In the first case, 10% density was desired, whereas the result was 17%. In the other case, 90% density was desired, whereas the result was 83%.

The `sprandsym()` function produces a symmetric matrix filled with random numbers in a sparse fashion. Since a symmetric matrix is a square matrix, defining only one dimension, say `n`, defines an *nxn* matrix.

```
1 >> sprandsym(5,0.2)
2 ans =
3
4 Compressed Column Sparse (rows = 5, cols = 5, nnz = 5 [20%])
5
6 (2,1) -> -0.39845
7 (3,1) -> -0.12342
8 (1,2) -> -0.39845
9 (2,2) -> -0.14327
10 (1,3) -> -0.12342
11
12 >> sprandsym(5,0.8)
13 ans =
14
15 Compressed Column Sparse (rows = 5, cols = 5, nnz = 20 [80%])
16
```

```
17  (1,1) -> -2.1227
18  (2,1) ->  1.5120
19  (3,1) -> -1.1275
20  (5,1) ->  1.0004
21  (1,2) ->  1.5120
22  (2,2) ->  0.28871
23  (3,2) ->  1.1516
24  (4,2) -> -0.11822
25  (5,2) ->  0.20848
26  (1,3) -> -1.1275
27  (2,3) ->  1.1516
28  (4,3) ->  0.20721
29  (2,4) -> -0.11822
30  (3,4) ->  0.20721
31  (4,4) ->  1.3381
32  (5,4) -> -1.3052
33  (1,5) ->  1.0004
34  (2,5) ->  0.20848
35  (4,5) -> -1.3052
36  (5,5) ->  0.64491
```

Using the spconvert function, you can define a sparse matrix with specific indices where you want to have non-zero elements. The first two columns represent the row and column indices, respectively, and the third and fourth columns represent the real and imaginary parts of the sparse matrix. The matrix can contain zero elements and the elements can be sorted in any order. Look at this example:

```
1  >> a = [1 2 3 4;1 3 4 4;1 2 3 0]
2  a =
3
```

```
4  1   2   3   4
5  1   3   4   4
6  1   2   3   0
7
8  >> spconvert(a)
9  ans =
10
11  Compressed Column Sparse (rows = 1, cols = 3, nnz = 2 [67%])
12
13  (1,2) -> 6 + 4i
14  (1,3) -> 4 + 4i
15
16  >> a = [1 2 3;1 3 4;1 2 3]
17  a =
18
19  1   2   3
20  1   3   4
21  1   2   3
22
23  >> spconvert(a)
24  ans =
25
26  Compressed Column Sparse (rows = 1, cols = 3, nnz = 2 [67%])
27
28  (1,2) -> 6
29  (1,3) -> 4
30
31  >> b=a'
```

```
32  b =
33
34  1   1   1
35  2   3   2
36  3   4   3
37
38  >> spconvert(b)
39  ans =
40
41  Compressed Column Sparse (rows = 3, cols = 4, nnz = 3 [25%])
42
43  (1,1) -> 1
44  (2,3) -> 2
45  (3,4) -> 3
46
47  >> a = [1 2 3 4;1 3 4 4;1 2 3 0]
48  a =
49
50  1   2   3   4
51  1   3   4   4
52  1   2   3   0
53
54  >> b=a'
55  b =
56
57  1   1   1
58  2   3   2
59  3   4   3
60  4   4   0
61
```

```
62  >> spconvert(b)
63  ans =
64
65  Compressed Column Sparse (rows = 4, cols = 4, nnz = 3 [19%])
66
67  (1,1) -> 1
68  (2,3) -> 2
69  (3,4) -> 3
```

5.6 Manipulating Arrays

Arrays can be manipulated in Octave by indexing them, creating new vectors, and slicing, flipping, sorting, and rotating them. The following sections cover these functions.

5.6.1 Indexing

Each element of the matrix is characterized by two numbers, the row number and the column number. This information is used to pinpoint an element and operate on it.

```
1  >> a = rand(2,3)
2  a =
3
4  0.5248873    0.5531882    0.0051345
5  0.1597312    0.3685503    0.3041072
6
7  >> a(2,3)=1
8  a =
9
```

```
10  0.5248873    0.5531882    0.0051345
11  0.1597312    0.3685503    1.0000000
12
13  >> a(1,1)=0
14  a =
15
16  0.00000    0.55319    0.00513
17  0.15973    0.36855    1.00000
```

Note that a(2,3)=1 sets the element in the second row and third column, i.e., number 0.3041072, to 1. Likewise, a(1,1)=0 sets the element in the first row and first column, i.e., number 0.5248873, to 0. To index numbers in a vector, you need a single number.

```
 1  >> a = [1,2,3,4,5,6,7,8,9]
 2  a =
 3
 4  1   2   3   4   5   6   7   8   9
 5
 6  >> a(1)
 7  ans =  1
 8  >> a(-1)
 9  error: subscript indices must be either positive integers
    less than 2^31 or logicals
10  >> a(5)
11  ans =  5
12  >> a(10)
13  error: A(I): index out of bounds ; value 10 out of bound 9
14  >>
```

It is important to note that, unlike some programming languages, where indices start at 0, in Octave indices start at 1. It will not take negative numbers as indices.

5.6.2 Using Indices to Make New Vector

```
1  >> a = [10 20 30 40 50 60]
2  a =
3
4  10    20    30    40    50    60
5
6  >> b = a ([1 3 6 1])
7  b =
8
9  10    30    60    10
```

In this example, b is a new vector formed from vector a where successive elements are made up of elements taken from the index vector [1 3 6 1].

```
 1  >> a = [11,12,13; 40,50,60; 17,18,19]
 2  a =
 3
 4  11    12    13
 5  40    50    60
 6  17    18    19
 7
 8  >> a([1,2], [2,3]) %row 1&2 as well as column 2&3
 9  ans =
10
11  12    13
12  50    60
```

Note that since use of comma operator is optional, henceforth we will define vectors and matrices by simply using whitespace.

5.6.3 Slicing

Matrices can be sliced to desired portions by using indices and the colon : operator.

```
1  >> a = [1 2 3 4 1 3 2 4 6 4 5]
2  a =
3
4  1   2   3   4   1   3   2   4   6   4   5
5
6  >> b =a(1:5)
7  b =
8
9  1   2   3   4   1
10
11 >> c = a(5:7)
12 c =
13
14 1   3   2
```

This is an important feature, as most experimental calculations demand filtering data. Here, a slice of data will be stored separately in a variable and then various mathematical operations can be performed on it. Now let's try to access slices of a multidimensional array. A matrix a is defined to be 5x5 matrix.

```
1  >> a = rand(5,5) % Defining a 5X5 matrix of random numbers
2  a =
3
4  0.563363    0.809636    0.910532    0.444515    0.425933
5  0.522041    0.926088    0.639679    0.972912    0.967932
6  0.842271    0.906763    0.272078    0.411484    0.337096
7  0.836302    0.320654    0.757441    0.459476    0.827371
8  0.305874    0.477885    0.175771    0.516654    0.039506
```

```
 9
10  >> b = a(1,1) % matrix with an element from row=1, column=1
11  b =  0.56336
12  >> c = a (1,:) % All elements of row=1
13  c =
14
15  0.56336    0.80964    0.91053    0.44451    0.42593
16
17  >> d = a(:,1) % All elements of column 1
18  d =
19
20  0.56336
21  0.52204
22  0.84227
23  0.83630
24  0.30587
25
26  >> e = a(:) % All elements of row and column as a column matrix
27  e =
28
29  0.563363
30  0.522041
31  0.842271
32  0.836302
33  0.305874
34  0.809636
35  0.926088
36  0.906763
37  0.320654
38  0.477885
39  0.910532
40  0.639679
```

```
41  0.272078
42  0.757441
43  0.175771
44  0.444515
45  0.972912
46  0.411484
47  0.459476
48  0.516654
49  0.425933
50  0.967932
51  0.337096
52  0.827371
53  0.039506
54
55  >> f = a(:,[1,3]) % all elements of column=1 and column=3
56  f =
57
58  0.56336     0.91053
59  0.52204     0.63968
60  0.84227     0.27208
61  0.83630     0.75744
62  0.30587     0.17577
63
64  >> g = a ([1,3],:) % all elements of row=1 and row=3
65  g =
66
67  0.56336     0.80964     0.91053     0.44451     0.42593
68  0.84227     0.90676     0.27208     0.41148     0.33710
```

- To access a single element, you use the index value of the row and column. For example, `b = a(1,1)` accesses the elements in the first row and first column.
- To access all elements of a row or column, you use the `:` operator. Hence, `c = a(1,:)` access all elements in the first row. Similarly, `>> d = a(:,1)` accesses all elements in the first column. A simple way to remember how to use the colon operator is that `:` stands for *all elements for.* Then you have the nth row/column where n is the given value.
- Using `a(:)`, you can create a new matrix, which is a column matrix having all the elements.
- A sub-matrix can be accessed by defining *all elements* for column/row and then defining indices in square brackets. For example, `f = a(:,[1,3])` defines a new matrix where elements are composed of *all elements of* the first and third columns. Similarly, `a([1,3],:)` defines all elements of the first and third rows.

You can compose complex sub-matrices using this powerful way of defining your choice of elements.

```
1  >> a = rand(5,6)
2  a =
3
4  0.1365941    0.7004691    0.4141496    0.1961403
   0.1386467    0.6338910
5  0.4073519    0.3970787    0.9404709    0.6876520
   0.6595586    0.1230414
```

```
 6  0.6775819     0.9203946     0.6048951     0.7997643
    0.6124899     0.2699103
 7  0.4513048     0.3531190     0.5228914     0.0504358
    0.6872609     0.3613488
 8  0.0071268     0.5250754     0.2268388     0.0047337
    0.2975212     0.3947907
 9
10  >> b = a([2,5],1:3)
11  b =
12
13  0.4073519     0.3970787     0.9404709
14  0.0071268     0.5250754     0.2268388
15
16  >> d = a([2,5],[1,3])
17  d =
18
19  0.4073519     0.9404709
20  0.0071268     0.2268388
21
22  >> e = a(2:5,1:3)
23  e =
24
25  0.4073519     0.3970787     0.9404709
26  0.6775819     0.9203946     0.6048951
27  0.4513048     0.3531190     0.5228914
28  0.0071268     0.5250754     0.2268388
```

This code defines a new 5x5 matrix called a and then defines a subset of this matrix using `a([2,5],1:3)`. This says, from *the second and third rows,* take elements from *the first column to the third column.*

Similarly `c = a(2:5,[1,3])` creates a matrix using this logic: *from the first* and *fifth* columns, take elements from *the second* and *third* rows. Now you can easily guess what `a([2,5],[1,3])` and `a(2:5,1:3)` should perform. It's a good idea to practice slicing arrays rigorously, as this is one of the most sought-after skills in data cleaning and data analysis in general.

5.6.4 Flipping a Matrix

The `flipud(A)` function returns a copy of matrix `A` with the order of the rows reversed. `flipud` stands for *flip up down*. `fliplr(A)` returns a copy of matrix `A` with the order of the rows reversed. `fliplr` stands for *flip left right*.

```
1  >> a = [1 2; 3 4; 5 6]
2  a =
3
4  1   2
5  3   4
6  5   6
7
8  >> fliplr(a)
9  ans =
10
11 2   1
12 4   3
13 6   5
14
15 >> flipud(a)
16 ans =
17
18 5   6
19 3   4
20 1   2
```

5.6.5 Rotating a Matrix

Using the `rot90(a,n)` command, you can rotate matrix `a` `n` times by 90 degrees.

```
1  >> a = [1 2; 3 4; 5 6]
2  a =
3
4  1   2
5  3   4
6  5   6
7
8  >> rot90(a,1)
9  ans =
10
11 2   4   6
12 1   3   5
13
14 >> rot90(a,2)
15 ans =
16
17 6   5
18 4   3
19 2   1
20
21 >> rot90(a,4)
22 ans =
23
24 1   2
25 3   4
26 5   6
```

5.6.6 Reshaping a Matrix

You can change the number of rows and columns in a matrix, provided that the total number of elements remains the same.

```
1  >> a = [1 2; 3 4; 5 6]
2  a =
3
4  1   2
5  3   4
6  5   6
7
8  >> reshape(a,6,1)
9  ans =
10
11 1
12 3
13 5
14 2
15 4
16 6
17 >> reshape(a,4,1)
18 error: reshape: can't reshape 3x2 array to 4x1 array
```

5.6.7 Sorting

You can sort numbers in increasing order using the `sort` function:

```
1  >> a = rand(1,5)
2  a =
3
4  0.577290    0.079980    0.880757    0.294744    0.964269
5
```

```
6 >> sort(a)
7 ans =
8
9 0.079980    0.294744    0.577290    0.880757    0.964269
```

5.7 Summary

This chapter illustrated various methods of auto-generating arrays in a desired fashion. Generating arrays for initialization enables you to write code more easily. Not having to define loops to fill up values in a matrix is a relief, because you can instead concentrate on auto-generating matrices rather than defining code for their generation.

The generation of random numbers inside matrices was also discussed. You can generate random numbers of desired distributions and they allow you to define *trial* data that matches reality closely. The ability to manipulate array dimensions as well as generate sub-arrays using slicing enables you to carve out smaller arrays from a bigger array based on designed rules that can be coded.

5.8 Bibliography

[1] http://www.math.sci.hiroshima-u.ac.jp/~m-mat/MT/emt.html

[2] http://www.jstatsoft.org/v05/i08/

CHAPTER 6

Plotting

6.1 Introduction

The ability to provide quality visualizations from output data is a key part of data analytics. Without visualization, numerical computations are difficult and sometimes impossible to interpret. Producing publication-quality images of complex plots that give a meaningful analysis of the numerical results was one of the biggest challenges for engineers and scientists all over the world when computers were introduced in the scientific domain. Many commercial software programs satisfied this need.

Octave provides this functionality too. Its plotting features include choosing from various types of plots in 2D and 3D formats, decorating plots with additional information such as titles, labeled axes, grids, and labels for data, and writing equations and other important information about data. The sections in this chapter describe these actions in detail. It is worth mentioning that plotting capabilities are essential to all simulation-based experiments since visual direction from the progressive steps give developers an intuitive understanding of the problem under consideration.

S. Nagar, *Introduction to Octave*, https://doi.org/10.1007/978-1-4842-3201-9_6

6.2 2D Plotting

Octave presents a host of built-in functions for generating a variety of 2D plots. These functions take an array as input arguments to define the data points and present plots in the desired fashion.

6.2.1 The plot(x,y) Function

Since you need to plot data on two axes, you first need to create them. Let's assume that the x axis has 100 linearly-spaced data points, whereby $y = x^2$. See Figure 6-1.

```
1  >> x = linspace(0,100,100);
2  >> y = x.^2
3  >> plot(x,y)
```

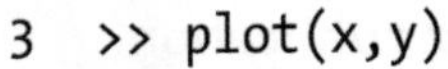

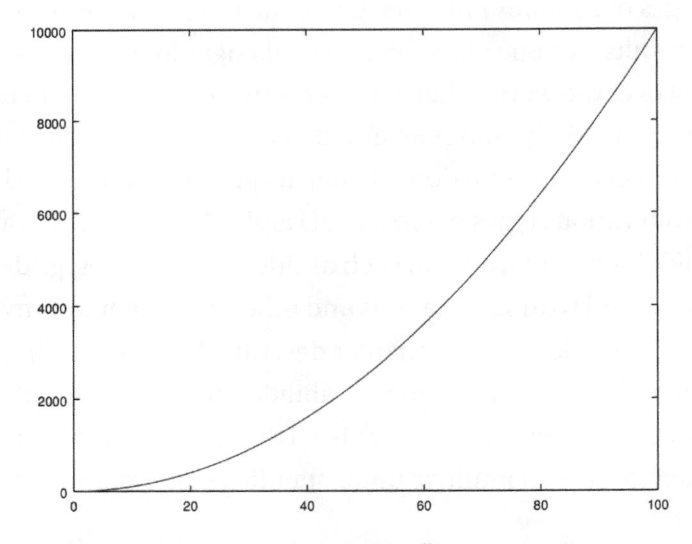

Figure 6-1. *Plotting $y = x^2$ with the plot() function*

First you define a variable x and place 100 equally-spaced data points from 0 to 100. This creates a 1*x*100 matrix. Using the scalar operation of exponentiation, you define the variable *y* as the x^2. Then you use the function `plot()`, which takes two arguments as the x-axis and y-axis data points. Typing `help(plot)` on the command prompt gives you useful insight into this wonderful function used to plot two-dimensional data.

6.2.2 The area() Function

The `area()` function results in a similar plot as the `plot()` function, but it also shades the area under the curve, as shown in Figure 6-2.

```
1  >> x = linspace(0,100,100);
2  >> y = x.^2;
3  >> area(x,y)
```

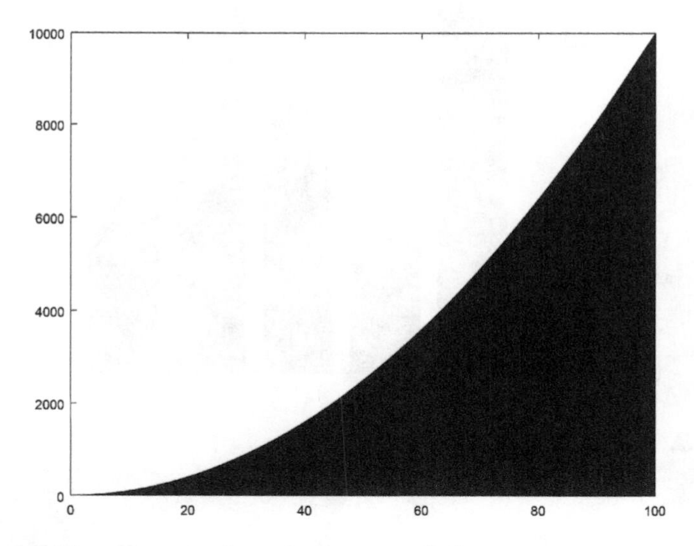

Figure 6-2. *Plotting $y = x^2$ with the area() function*

6.2.3 The bar(), barh(), and hist() Functions

Bar charts are primitive but very effective in visualizing primary statistical information. There are three ways to plot bar charts and histograms:

- `bar()` plots a vertical bar chart (see Figure 6-3)
- `barh()` plots a horizontal bar chart (see Figure 6-4)
- `hist()` plots a histogram chart (see Figure 6-5)

```
1  >> x = [1,2,3,4,5,6];
2  >> y = [0.5,2.2,0.7,1.5,2.5,0.9];
3  >> bar(x,y)
```

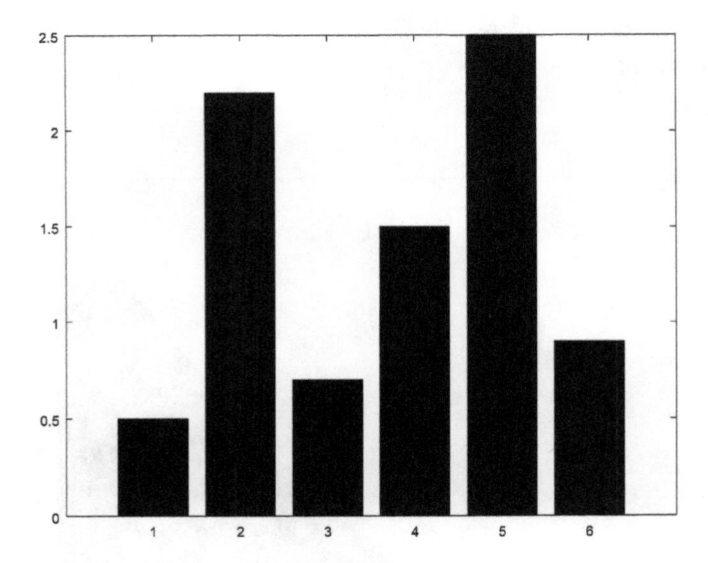

Figure 6-3. *A bar chart*

In a similar fashion, a horizontal bar chart can be plotted using the barh() function.

```
1  >> x = [1,2,3,4,5,6];
2  >> y = [0.5,2.2,0.7,1.5,2.5,0.9];
3  >> barh(x,y)
```

A histogram can be plotted using the hist() function as well. Let's next check the behavior of a normalized distribution of random numbers generated by the randn() function (see Figure 6-5).

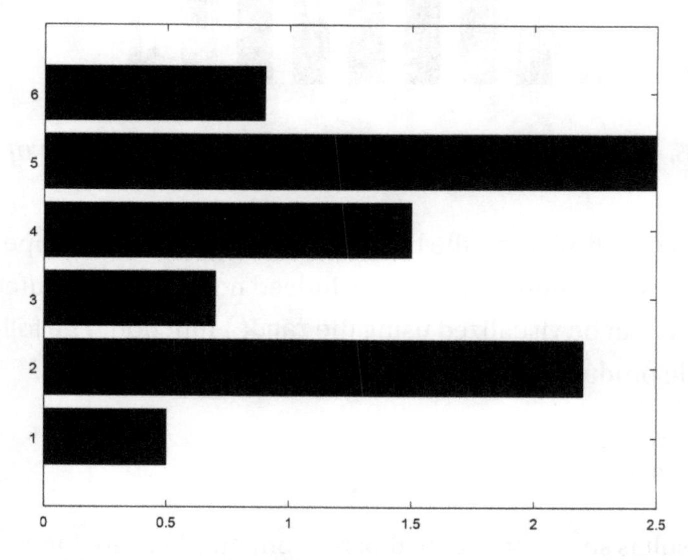

Figure 6-4. *A horizontal bar chart*

```
1  >> x = randn(100);
2  >> hist(x)
```

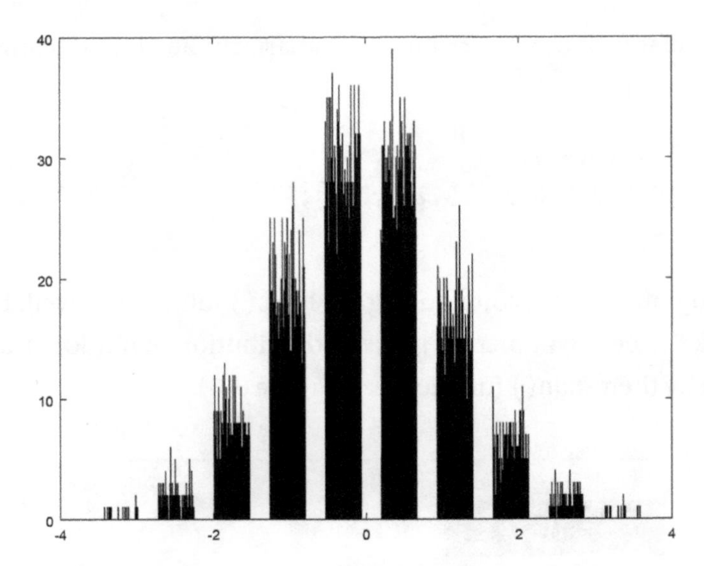

Figure 6-5. *A histogram showing normalized distribution of random numbers*

You can clearly observe the bell-shaped curve of the envelope to confirm that the random numbers are indeed normally distributed. Other distributions can be visualized using the `rand()` function. The following Octave code produces uniformly distributed random numbers.

```
1  >> x = rand(100);
2  >> hist(x)
```

The result is shown in Figure 6-6; random numbers are indeed found to be uniformly distributed over the range (1, 2).

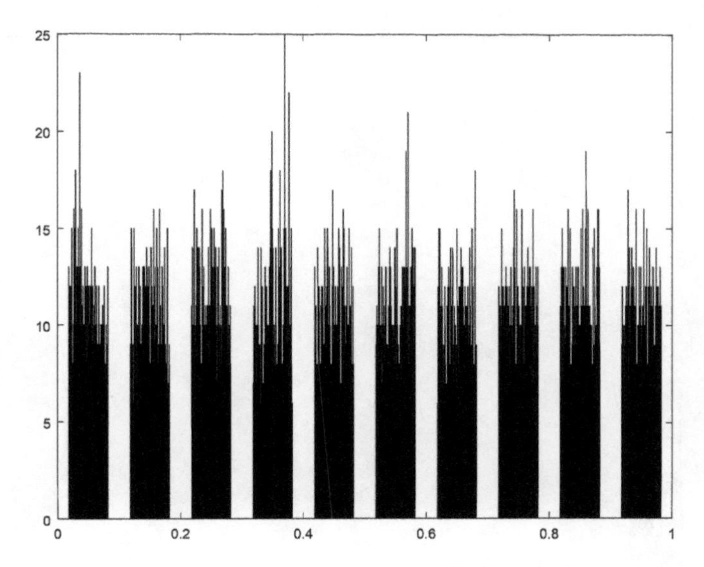

Figure 6-6. *A histogram showing uniform distribution of random numbers*

In a similar manner, exponentially distributed random numbers can be visualized using the `rande()` function. The following Octave code produces exponentially distributed random numbers.

```
1  >> x = rande(100);
2  >> hist(x)
```

The result is shown in Figure 6-7; random numbers are indeed found to be uniformly distributed over the range (1, 2).

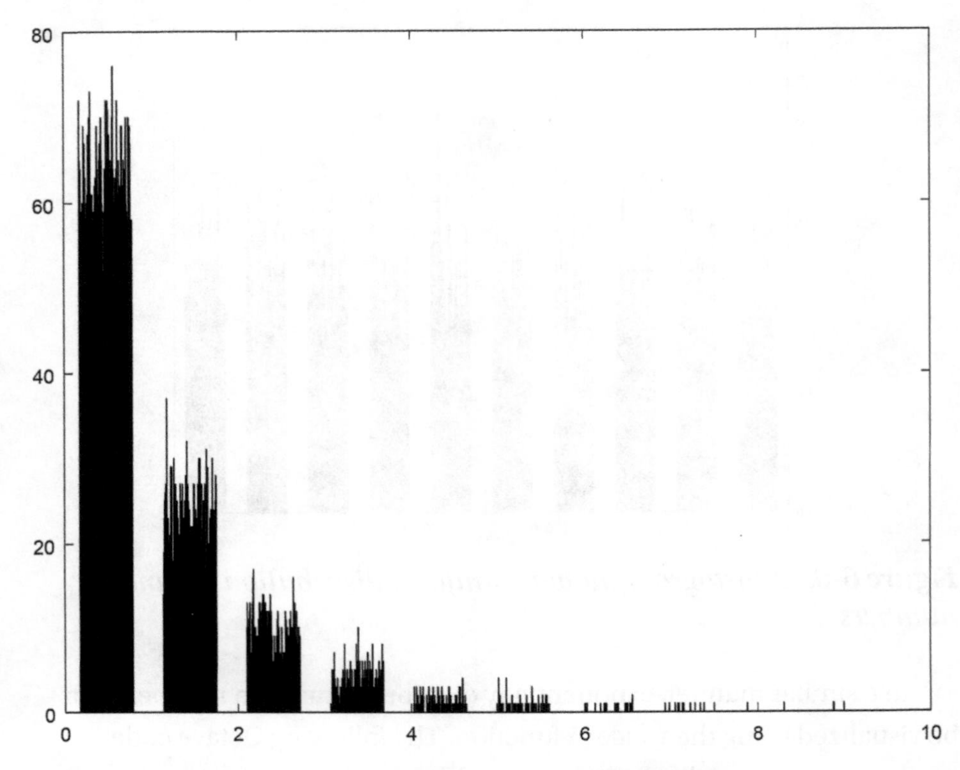

Figure 6-7. *A histogram showing exponential distribution of random numbers*

Plotting Multiple Plots on the Same Graph

Multiple plots can be plotted on the same graph by simply supplying *x* and *y* axes vectors, as shown in Listing 6-1.

Listing 6-1. The multi.m File

```
1  clear all;
2  clf;
3  x = linspace(1,100,100);
4  y1 = x.^2.0;
5  y2 = x.^2.1;
```

```
6  y3 = x.^2.2;
7  y4 = x.^2.3;
8  plot(x,y1,"@12",x,y2,x,y3,"4",x,y4,"+")
9  grid on
10 legend('x^2','x^{2.1}','x^{2.2}','x^{2.3}');
11 xlabel('x-axis')
12 ylabel('y-axis')
13 title('Multiple Graphs')
14
15 %plot y with points of type 2 (displayed as '+')
16 %and color 1 (red), y2 with lines, y3 with lines
17 %of color 4 (magenta) and y4 with points displayed as '+'
```

Explanation of the lines in Listing 6-1 follows:

1. `clear all` clears the variable names and values from memory.
2. `clf` clears any current window.
3. `x = linspace(1,100,100)` creates a vector called `x` made up of 100 equally spaced data points between 1 and 100.
4. $y1 = x.^{2.0}$, makes a new vector named `y1` having an element-wise square of vector `x`.
5. $Y2 = x.^{2.1}$, makes a new vector named `y2` having an element-wise exponentiation by 2.1 of vector `x`.
6. $Y3 = x.^{2.2}$, makes a new vector named `y3` having an element-wise exponentiation by 2.2 of vector `x`.
7. $Y4 = x.^{2.3}$, makes a new vector named `y4` having an element-wise exponentiation by 2.3 of vector `x`.
8. Plots the numbers as per comment given in lines 15,16,17.

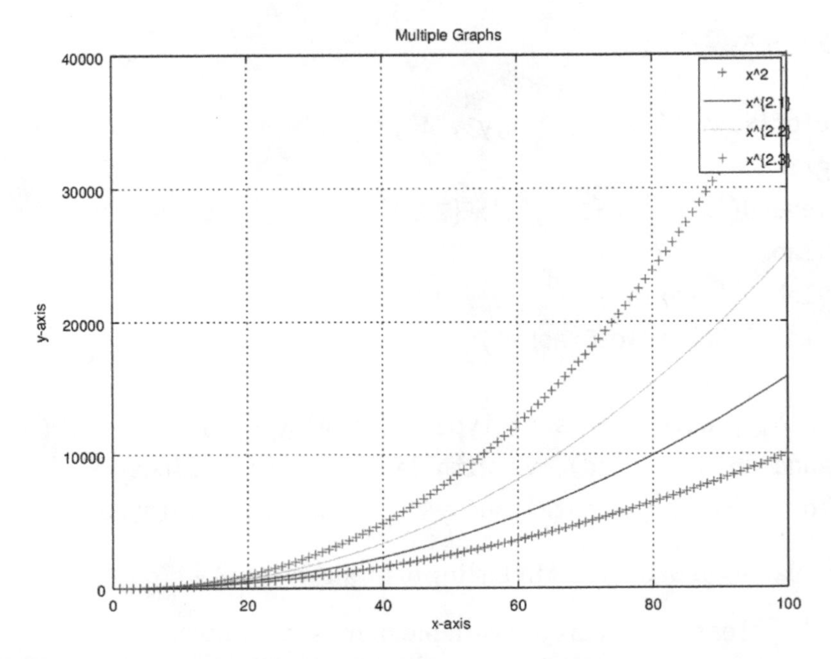

Figure 6-8. *Multiple plots within the same figure*

9. Grid is turned on for the figure.
10. `xlabel` takes the value of the string `x-axis` in line 11.
11. `ylabel` takes the value of the string `y-axis` in line 12.
12. `title` takes the value of the string `Multiple Graphs` in line 13.

Figure 6-8 is obtained by running the code. These types of plots are used to check the variation of results by varying a particular parameter.

Plotting Multiple Plots Separately

The `subplot(row,column,index)` command is used to plot multiple plots within the same figure separately. `subplot(2,2,4)` means that the plot will be on the second row, in the second column, and in the fourth index. See Listing 6-2.

Listing 6-2. The multiSubplot.m File

```
1  clear all;
2  clf;
3  x = linspace(1,100,100);
4  y1 = x.^2.0;
5  y2 = log(x);
6  y3 = sin(x);
7  y4 = log10(x);
8  subplot(2,2,1), plot(x,y1)
9  subplot(2,2,2), plot(x,y2)
10 subplot(2,2,3), plot(x,y3)
11 subplot(2,2,4), plot(x,y4)
12 %gridon
13 %legend('x^2','x^{2.1}','x^{2.2}','x^{2.3}');
14 %xlabel('x-axis')
15 %ylabel('y-axis')
16 %title('Multiple Graphs')
17
18 %plot y with points of type 2 (displayed as '+')
19 %and color 1 (red), y2 with lines, y3 with lines
20 %of color 4 (magenta) and y4 with points displayed as '+'
```

As shown in Figure 6-9, plots are organized as matrices, where row numbers and column numbers dictate the position. An index of the plot can then be used to treat it as an object for further processing on the graphical objects.

You can learn about other commands for controlling the font size, tick labels, fonts, mathematical equations, etc. by typing `help plot` or reading the documentation of this function. There are also many examples on the web. You will use this function frequently, so make sure you have good command over its use.

6.2.4 Plotting in Polar Coordinates

Sometimes you might prefer to plot in polar coordinates, rather than in Cartesian coordinates. Then, instead of using *x*, *y*, the coordinates are *r*, θ. See Listing 6-3.

Listing 6-3. The CoordinatesPolar.m File

```
1  theta = 0:0.02:2*pi;
2  a1 = 0.5 + 1.3.^theta;
3  a2 = 5*cos(theta);
4  a3 = 3*(1 - cos(theta));
5  a4 = 6*sin(4*theta);
6  r = [a1;a2;a3;a4];
7  PolarGraph = polar(theta,r,"*");
8  set(PolarGraph,"LineWidth",2);
9  legend("spiral","circle","heart","Rose");
```

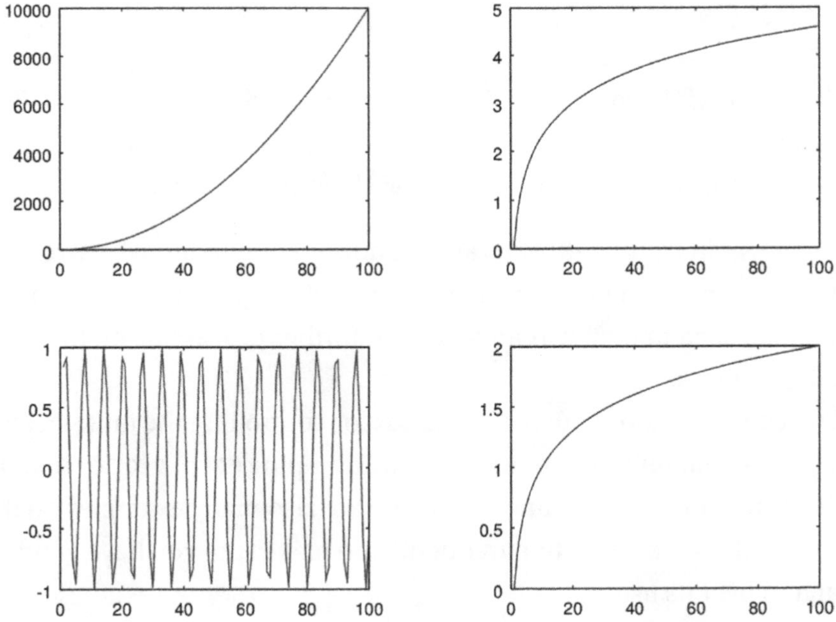

Figure 6-9. *Separate multiple plots within the same figure*

Figure 6-10 shows an example of a polar graph for code given by the `CoordinatesPolar.m` example. Explanation of this program follows (according to line number):

1. A variable named `theta` representing θ is defined by points starting from 0 to 2π with steps of 0.02.

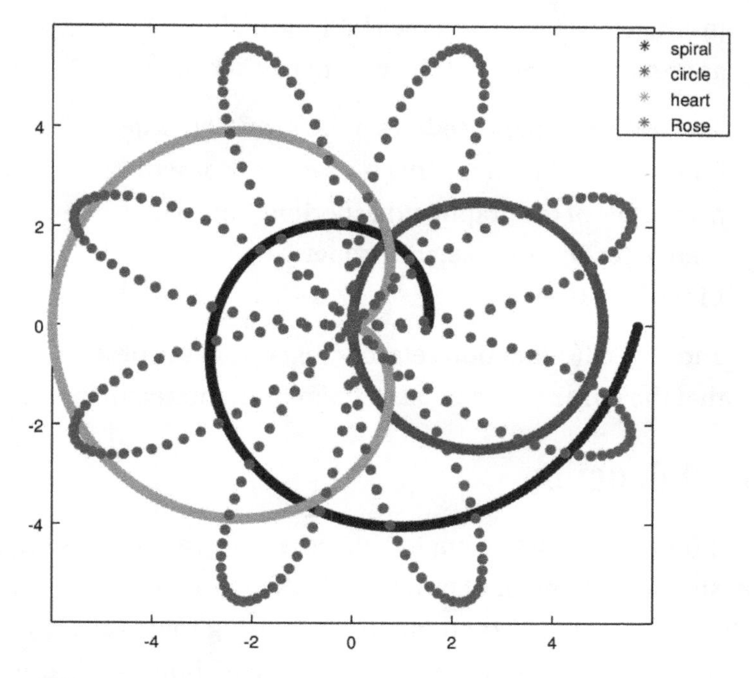

Figure 6-10. *A polar graph*

2. A variable named `a1` representing *r* for **spiral** is calculated by equation $r = 1.5(\theta)$.

3. A variable named `a2` representing *r* for **circle** is calculated by equation $r = 5(cos(\theta))$.

4. A variable named `a3` representing *r* for **heart** is calculated by equation $r = 3(1\text{-}cos(\theta))$.

5. A variable named `a4` representing r for **rose** is calculated by equation $r = 6(sin(4\theta))$.

6. A variable named `r` stores all the *rs* calculated using the equations as a column vector.

7. A variable named `PolarGraph` stores the values produced by the function `polar()`, which takes θ, r as arguments and * for the type of marker.

8. The `set` function is used to set the *property values* for the graph function. This is a neat way of setting properties of the graph and experimenting with them later. In this case, the property named `LineWidth` is set to `2`.

9. The `legend()` function sets four legends in the order that the `polar` function takes them from the vector `r`.

The rose() Function

The `rose()` function draws an angled histogram, i.e., a polar histogram. The input should be a vector of numbers. Let's look at its use by constructing a vector of 100 random numbers using `randn(100,1)*pi` and then feeding it to the `rose()` function. The result is shown in Figure 6-11.

```
1  >> x = randn(50,1)*pi/2;
2  >> rose(x)
```

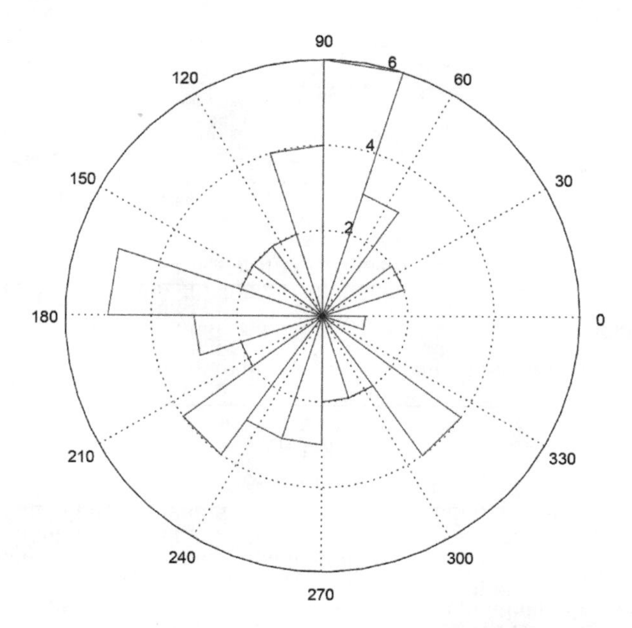

Figure 6-11. *Plot of random numbers by the rose() function*

6.2.5 Logarithmic Plots

For plotting graphs involving logarithmic scale, MATLAB provides three options:

- `semilogx()`: Plots with logarithmically spaced x-axis. As an example, consider the `log1a.m` file shown in Listing 6-4, which produces the plot shown in Figure 6-12.

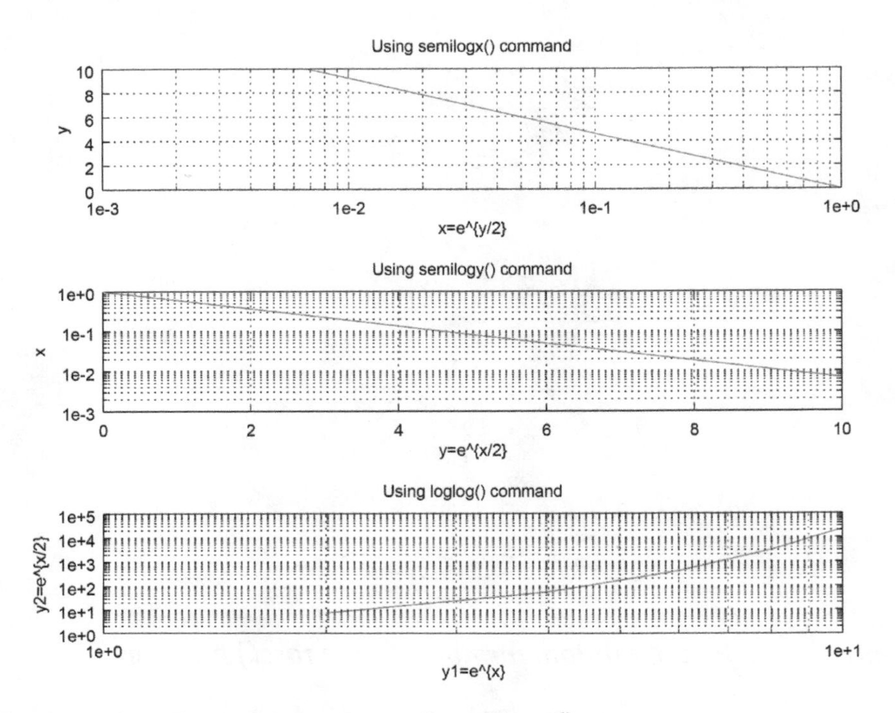

Figure 6-12. *Describing usage of semilogx()*

Listing 6-4. The log1a.m File

```
 1  %Octave program to illustrate
 2  %usage of semilogx() and
 3  %semilogy() and loglog() command
 4
 5  %semilogx()
 6  y = 0:2:10;
 7  x = exp(-y/2);
 8  subplot(3,1,1)
 9  semilogx(x,y)
10  grid on
11  xlabel('x=e^{y/2}');
12  ylabel('y');
```

```
13  title('Using semilogx() command');
14
15  %semilogy()
16  x1 = 0:2:10;
17  y1 = exp(-x1/2);
18  subplot(3,1,2)
19  semilogy(x1,y1)
20  grid on
21  xlabel('y=e^{x/2}');
22  ylabel('x');
23  title('Using semilogy() command');
24
25  %loglog()
26  x2 = 0:2:10;
27  y1 = exp(x2);
28  y2 = exp(x2/2);
29  subplot(3,1,3)
30  loglog(x1,y1)
31  grid on
32  xlabel('y1=e^{x}');
33  ylabel('y2=e^{x/2}');
34  title('Using loglog() command');
```

- `semilogy()`: Plots with a logarithmically spaced y-axis.
- `loglog()`: Plots with both axes logarithmically spaced.

The pie() Function

You can create a pie chart using the `pie()` function. This provides a very powerful tool to visualize the parts of a whole. The usage is explained in the following code and the images are shown in Figure 6-13. The `pie()` function supports 34 items such that `a,b,c,d,e,f` get `4,7,2,8,4` and `9` parts.

The pie chart can be made by first defining the parts as an array, then defining the labels as an array. Then if `pie()` function is fed directly, you'll get a color-coded exploded pie chart showing the percentages of each part. When a `show()` array is also used, it explodes only those parts whose corresponding element is 1.

```
1  >> x = [3,2,1,4,1,2];
2  >> subplot(2,1,2)
3  >> pie(x)
4  >> subplot(2,1,1)
5  >> show = [0,1,0,1,0,1];
6  >> pie(x,show,labels)
```

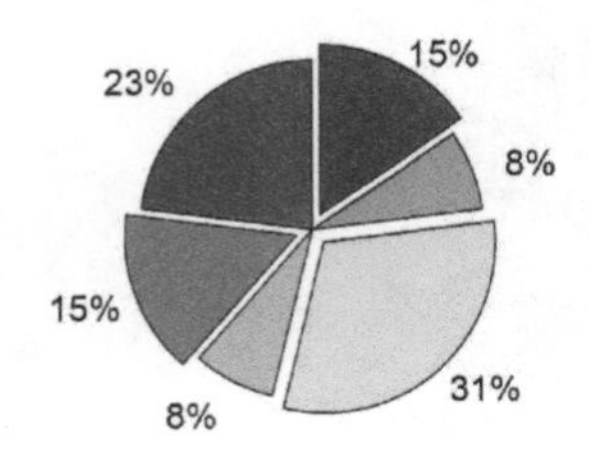

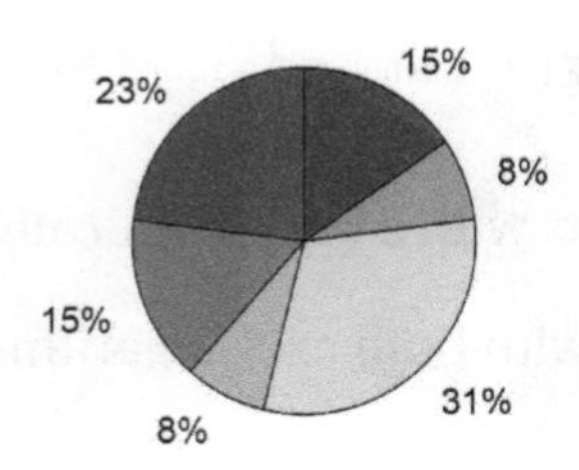

Figure 6-13. *Pie plots with all parts exploded and with some parts exploded*

The stairs() Function

A `stairs()` function draws a stair-step graph for elements of a vector. Consider an example of plotting $y = sin(x)$, where x is a vector of 100 elements from 0 to 20. As you can see in Figure 6-14, the sinusoidal wave can be visualized whereby the data points are connected in a stair-step fashion.

```
1  >> x = 0:20:1000;
2  >> y = sin(x);
3  >> stairs(y)
```

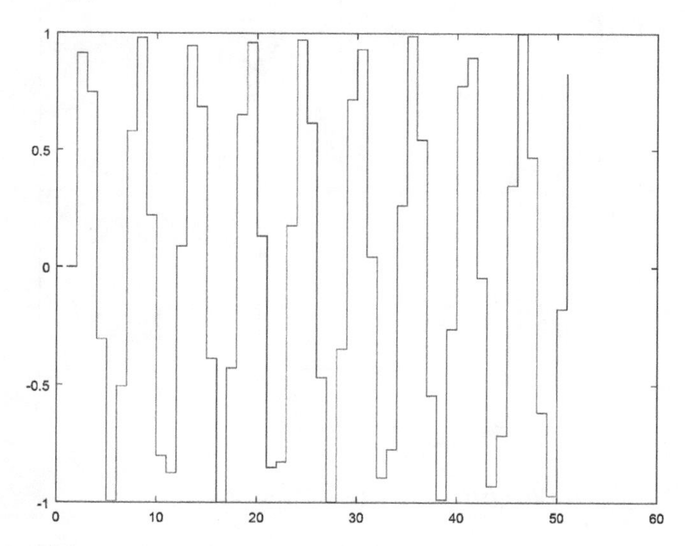

Figure 6-14. *A stair plot for* $y = x^{2.5}$

The stem() Function

Stem plots draw data points as stems that extend from equally spaced values. The code that plots $y = sin(x) \in (-4\pi, 4\pi)$ is shown here. It will produce the graph shown in Figure 6-15.

```
1  >> x = -4*pi:4*pi;
2  >> y = sin(x);
3  >> stem(y)
```

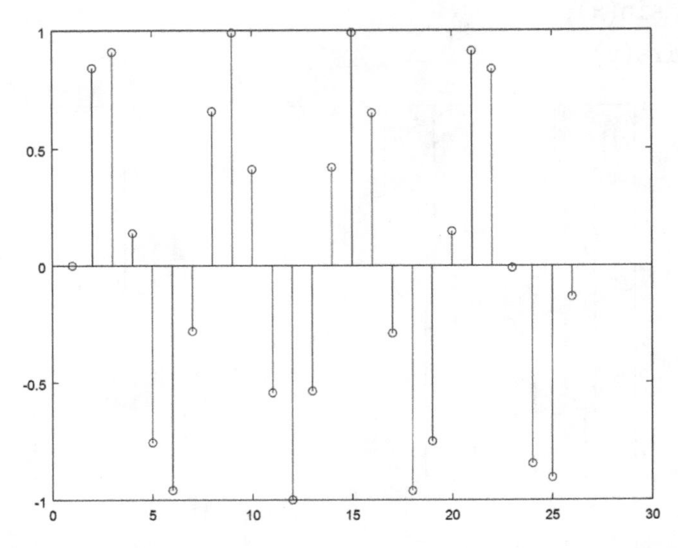

Figure 6-15. *Stem plot for $y = cos(x) \in (-\pi, \pi)$*

6.2.6 Creating 3D Plots

Octave has various functions available for 3D plotting. Choosing the best one depends on your particular problem.

The mesh Function

Listing 6-5 shows the `mesh` function in action. This code will produce the 3D graph shown in Figure 6-16.

Listing 6-5. The ThreeDMesh.m File

```
1  a = b = linspace(-8,8,41)';
2  [xx,yy] = meshgrid(a,b);
3  c = sqrt(xx.^2+yy.^2) + eps;
4  d = sin(c)./c;
5  mesh(a,b,d);
```

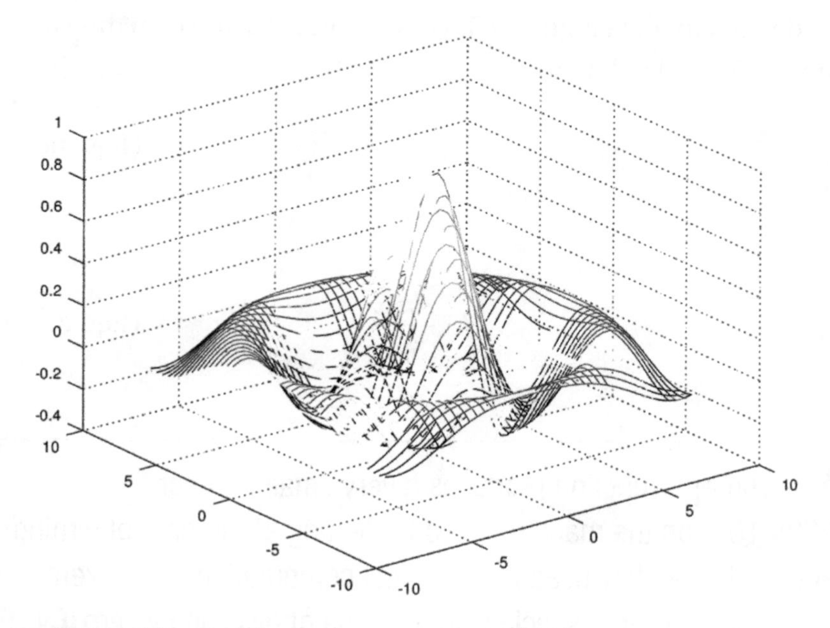

Figure 6-16. *3D meshing*

It's important to note that this code uses a new function named `meshgrid`. Use `help meshgrid` to learn a bit about this function. It is used as follows:

```
1  >> a = b = linspace(-8,8,41);
2  >> [xx,yy] = meshgrid(a,b);
```

Two variables are created—a and b—and they store 41 linearly spaced data points between -8 and 8 as a row vector. These two row vectors (both are 1*X*41 in dimension) are passed as arguments for the `meshgrid` function, which gives two outputs: xx and yy. These are 41*X*41 dimensioned matrices, where rows of xx are copies of a and columns of yy are copies of b. The `meshgrid` function can also take a third argument to create a complete 3D grid. Otherwise, on the two-dimensional base grid, a function can be defined for data points defined by copies of a and b. In this case, the function is defined as follows:

$$c = \sqrt{x^2 + y^2} \qquad \text{(Equation 6-1)}$$

and

$$d = \frac{sin(c)}{c} \qquad \text{(Equation 6-2)}$$

Note The `eps` function produces a very small number ($2.2204.10^{-16}$ on the machine used for testing at the time of writing the book). It is widely used in numerical computation where zero needs to be avoided, especially in the case of division by zero. By adding a very small number to large numbers, you avoid this problem (remember that variable c calculated in Step 3 is then used in division as the denominator in Step 4). (see Listing 6-5)

Continuing now with the plotting exercise, new arrays can then be used to plot by applying the 3D plotting function `mesh()`, which takes two arrays a and d as its arguments. This results in the graph in Figure 6-16.

If `mesh(x,y,z)` is used, then a wireframe mesh made up of rectangles is created. The vertices of the rectangles are made up of data points generated by the function (in this case, Equations 6-1 and 6-2). The (x, y) coordination of vertices is given by the xx and yy matrices, since the x coordinate comes from the xx matrix and the y coordinate comes from the yy matrix. z determines the height above the plane of each vertex. In this way a 3D plot is plotted.

It is important to note that the original 3D curve is interpreted as a surface made of flat rectangles, which is, at best, an approximation. In some cases, this error can be ignored. To get less error, make the rectangles smaller, if possible. There are other variations of this same function, such as `ezmesh`, `meshc`, and `meshz`. A simple `help` command can be very useful in determining which one is best for a particular problem.

The mesh also color-codes for height (the z-value). This is computed by linearly scaling the z values to fit the range of the current color-map (use `help colormap` to learn more).

The meshc Function

`meshc()` generates a 3D rectangular mesh as well as contour at the base. As shown in Figure 6-17, apart from producing a 3D plot for a given function, you also obtain a contour plot. Note that this time, the equation is working on matrices and is written as an argument of the `meshc()` function, which makes the programs even smaller. See Listing 6-6.

Listing 6-6. The ThreeDMeshc.m File

```
1  x = linspace(-10,10,50);
2  y = linspace(-10,10,50);
3  [xx,yy] = meshgrid(x,y);
4  meshc(xx,yy,2-(xx.^2+yy.^2))
```

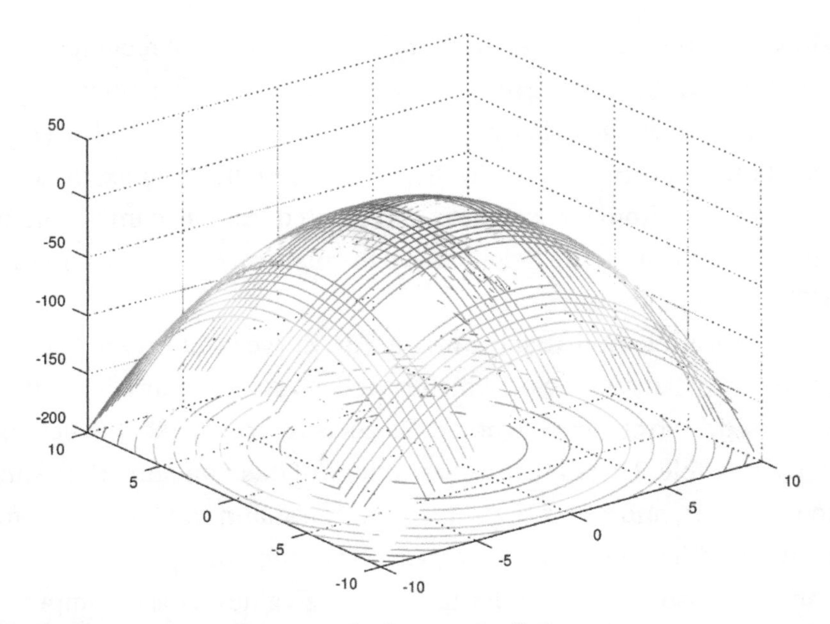

Figure 6-17. *3D meshing with the meshc() function*

The surf() Function

The `surf()` function (see Listing 6-7) generates a surface plot whereby a wire mesh is simply filled up at the empty points, as shown in Figure 6-18.

Listing 6-7. The ThreeDsurf.m File

```
1  a = b = linspace(-8,8,10)';
2  [xx,yy] = meshgrid(a,b);
3  c = sqrt(xx.^2 + yy.^2) + eps;
4  d = sin(c)./c;
5  surf(c,d);
```

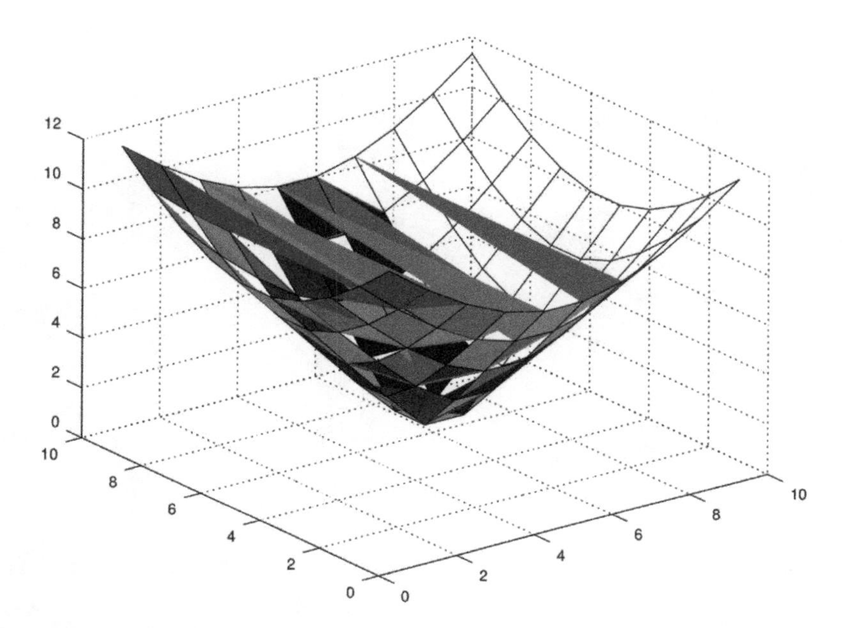

***Figure 6-18.** 3D meshing with the surf() function*

6.3 Summary

A rich library of plotting functions makes Octave a suitable choice for plotting data in a variety of publication-ready formats. Together with commands to access systems files and folders, these plots can be directed to be saved at appropriate places for making a suitable report. Plotting in 3D and viewing with different angles is quite intuitive in Octave. Hence, Octave is a suitable choice to visualize your data.

CHAPTER 7

Data Through File Reading and Writing

7.1 Introduction

Once you have mastered the art of defining and manipulating arrays and plotting the output, you can formulate physical problems in terms of numerical computations and solve them on a digital computer. This process has some requirements:

- The data should be in digital form (a digital file).
- The computer program should be able to read the file and create arrays without errors. If errors have been made, then a mechanism to check those errors and give a warning should be in place. If possible, you should also be able to correct them.
- The data should be stored as an array in the proper data type and should be displayed on demand in the proper format.
- Array operations on data will result in memory usage in terms of reading and writing data on-disk. This should be facilitated by the system. The user should be able to check the status of memory as and when required.

S. Nagar, *Introduction to Octave*, https://doi.org/10.1007/978-1-4842-3201-9_7

- Post-processing tasks include displaying data in various formats. This includes as a printout from the printer, on a terminal, as a graph on the terminal or printer/plotter, etc.
- Generate a report, file, or graph based on the data of a particular experiment if possible, as this makes the user's tasks easier.

Octave includes features related to each of these steps. This chapter discusses them briefly.

7.2 File Operations

File operations are an important part of computation. It is important to note that the file system is OS-dependent. Octave was traditionally written for UNIX-like systems, so it works on Linux-based and Mac OSX equally well and with the same set of commands. On Windows, you'll use the same commands as in Linux when dealing with files. The code in this chapter was tested on Windows 8, Mac OSX 10.10, and Ubuntu 14.04 systems.

7.2.1 Users

A computing system is accessed by many different users. Each user creates a workspace to avoid damaging each other's work. After login, a user's workspace becomes active for that user. The workspace is made up of various files and folders. Some files are essential for the OS to define the workspace and its properties, hence they should not be altered . This is ensured by assigning permissions to various users. For example, reading and writing a file are both restricted by permissions. The administrator (called the `admin`) is also called the "super user" and has permission to edit any file/folder. You must understand the defined user type for the computer system and then issue commands accordingly. If you are not permitted to access certain folders and then data is placed inside those files/folders, you will be denied access (unless you ask the admin to change your permissions).

7.2.2 File Path

A directory/folder can contain subdirectories/subfolders and files again. This can go to any level if this process if not restricted by the administrator.

The pwd command stands for *print working directory*. On an Octave terminal, typing pwd displays the path of the present working directory, as shown here:

```
1  >> pwd
2  ans = /home/sandeep
```

In the user's /home directory, there is another directory named /sandeep. This is the present working space. When pwd is typed on the terminal, a variable named ans stores this data (the file path). A variable name of your choice can be assigned to store the filename as a string.

A file/folder is accessed by typing the file path at the terminal. Let's do a small exercise to understand this process. To create a new directory, use the mkdir name command as follows:

```
1  >> mkdir octave
2  ans = 1
3  >> ls
4  Downloads                        Music
5  R
6  Templates
7  octave
8  Videos
9  Desktop                          software
```

```
10 Work
11 Documents            Library
12 Pictures
13 >> cd octave
14 >>
```

On line 1, mkdir octave creates a directory named octave. To see the contents of the present directory, you can use the ls command, as shown on line 3, which stands for *list*. To change the directory, you can use the cd file path command, as shown in line 13, which is changing to the octave directory in this case. I suggest that you work in this directory for the rest of the book.

7.2.3 Creating and Saving Files

The save and load commands enable you to write and read data to memory.

```
 1 >> matrix = rand(3,3);
 2 >> save MyFirstFile.mat matrix
 3 >> ls
 4 MyFirstFile.mat
 5 >> load MyFirstFile.mat
 6 >> matrix
 7 matrix =
 8
 9    0.467414   0.610273   0.429941
10    0.568490   0.037898   0.734682
11    0.547370   0.275421   0.539650
12
13 >>
```

On line 1, a variable named matrix is created. It stores a 3x3 matrix with random values. On line 2, this data is stored as a .mat file named MrFirstFile.mat, which is passed the variable name as the argument. When required, this file can be loaded in the workspace using the load MyFirstFile.mat command and then calling the variable name matrix. The random numbers recorded at the time of saving the file are loaded into the 3x3 matrix. Note that the data doesn't have to be numbers. It can be anything that a digital computer can handle, including pictures, videos, strings, and characters, just to name a few.

Multiple variables can be stored in the same file by passing the name of the variables at the time of saving.

```
 1  >> matrix1 = rand(4,4);
 2  >> matrix2 = rand(2,3);
 3  >> matrix3 = rand(2,2);
 4  >> save ("SavingMultipleVariables.mat","matrix1","matrix2",
       "matrix3")
 5  >> load SavingMultipleVariables.mat
 6  >> matrix1
 7  matrix1 =
 8
 9     0.8598130   0.0118250   0.9803720   0.3044413
10     0.6676748   0.0056845   0.1101545   0.2183920
11     0.2547204   0.8192626   0.8056112   0.6961116
12     0.7924558   0.9130480   0.1976146   0.4635055
13
14  >> matrix2
15  matrix2 =
16
17     0.35215   0.55770   0.66650
18     0.98515   0.98677   0.45513
19
```

```
20  >> matrix3
21  matrix3 =
22
23     0.097693   0.540354
24     0.923853   0.329501
25
26  >>>> save -binary SavedAsBinary m*
27  >> ls
28  MyFirstFile.mat SavedAsBinary   SavingMultipleVariables.mat
```

The help save and help load commands give very useful instructions about using these features. Using options, you can save the file in a specific format. For example, on line 26, all variables names starting with m are saved as binary data inside a binary file named SavedAsBinary. This is particularly important when data generated from Octave-based numerical computation is used to feed other programs. You can also specify the precision of saved data using options. You can also compress a big file using a -zip command. This is very useful when the data generated by Octave is large and needs to be transmitted.

The load function follows the same logic as the save function. Data can be unzipped and loaded from a particular formatted file as an array. The array, thus populated, can be used for computation and the resultant files can be created using the save function again (if required). Elaborate computations require this procedure to be repeated successively many times, so the functions have been optimized to locate and load the required data quickly.

The diary Command

An Octave session can be recorded in a file by using the diary command. Use help diary to look at its usage in detail. Typing help filename allows you to record the session in a file with a given filename. The commands and their outputs are continuously updated using this function.

Using the history command, a list of executed commands is displayed. Various options are available to see this history in its particular formats.

Opening and Closing Files

To read and write data files, they must be opened and defined as readable and/or writable. The fopen function returns a pointer to an open file that is ready to be read or written. This is defined by an option r as readable, w as writable, r+ as readable and writable, a for appending (writing) new content at the end of the file, and a+ for reading, writing, and appending. The opening mode can be set as t for text mode or b for binary mode. z enables you to open a zipped file for reading and writing.

Once all the data has been read from or written to the opened file, it should be closed. The fclose function does this.

```
1  MyFile = fopen("a.dat","r");
```

A variable called MyFile is created and is used to store the contents of the a.dat file. This file is opened in reading mode, which means it cannot be edited. This is important if you want the file to remain unchanged while sharing the information with others. freport() prints a list of files opened and whether they are opened for reading, writing, or both. For example:

```
1  >> freport
2
3  number   mode   arch       name
4  -------  -----  -----      ----
5  0        r      ieee-le    stdin
6  1        w      ieee-le    stdout
7  2        w      ieee-le    stderr
8
9  >>
```

Reading and Writing Binary Files

A binary file is a computer-readable file. They are simply sequences of bytes. Same as C functions, the `fread` and `fwrite` functions can read and write binary data from a file.

The csvread and csvwrite Functions

The `csvread` and `csvwrite` functions are used to read data from `.csv` files, which stands for *comma separated values.*

Suppose the following data needs to be stored as a `.csv` file.

```
1  2  3  4
5  6  7  8
8  7  6  5
4  3  2  1
```

The following code creates an array using `csvwrite` to create a file named `csvTestData.dat` containing the matrix values. You can check by simply opening this newly created file in a text editor. On line 3, a new file named `csvTestData1.dat` is created with an offset defined at row 1 and column 2.

```
 1  >> a = [1,2,3,4;5,6,7,8;8,7,6,5;4,3,2,1];
 2  >> a
 3  a =
 4
 5  1  2  3  4
 6  5  6  7  8
 7  8  7  6  5
 8  4  3  2  1
 9  >> csvwrite('csvTestData.dat',a)
10  >> csvwrite('csvTestData1.dat',a,1,2)
11  >> a1 = csvread('csvTestData.dat')
```

```
12  a1 =
13
14  1   2   3   4
15  5   6   7   8
16  8   7   6   5
17  4   3   2   1
18
19  >> a1 = csvread('csvTestData.dat',1,2)
20  a1 =
21
22  7   8
23  6   5
24  2   1
25
26  >>
```

Now, the csvread function can be used to create matrices with the desired offsets just as the function csvwrite.

Note A number of other functions that read and write files exist, but the present section focuses on some of the most commonly used ones. Access the documentation to learn more about using these specialized functions, if required.

7.2.4 Working with Excel Files

A lot of data is present on the Internet in the form of Excel files. Octave has a separate module to work with these files, but it first needs to be installed. The module IO is part of the octave-forge project. To install a module, you have to type pkg install -forge package_name at the Octave command prompt:

```
1  pkg install -forge io
```

Note that you must be connected to the Internet in this case.

Once the module has been automatically installed in the proper place, you can use its functions. The following is the list of file extensions and associated permissions.

```
1  File extension          COM  POI  POI/OOXML  JXL  OXS  UNO  OTK  JOD  OCT
2  --------------------------------------------------------------------------
3  .xls (Excel95)          R                    R         R
4  .xls (Excel97-2003)     +    +    +          +    +    +
5  .xlsx (Excel2007+)      ~         +               (+)  +              +
6  .xlsb, .xlsm            ~                         ?    R              R?
7  .wk1                    +                              R
8  .wks                    +                              R
9  .dbf                    +                              +
10 .ods                    ~                              +    +    +    +
11 .sxc                                                   +         +
12 .fods                                                  +
13 .uos                                                   +
14 .dif                    +                              +
15 .csv                    +                              R
16 .gnumeric                                                             +
17 --------------------------------------------------------------------------
18
19 R  :  only read; + :  full read/write; ~ : dependent on Excel version
```

Working with Excel Files

The `xlsopen`, `xlswrite`, `xlsclose`, `odsopen`, `odswrite`, and `odsclose` commands open, write, and close `.xls` and `.ods` files. While `.xls` files are generated using Microsoft Excel software, `.ods` files are generated using Open/Libre Office software, which is the open source equivalent of Microsoft Excel. The process of opening, reading, and writing data is as follows:

- `xlsopen('Filename.xls')`
- `a = xlsread('Filename.xls', '3rd_sheet', 'B3:AA10');`

  The numeric data from the `Filename.xls` worksheet's `3rd sheet` will be read from cell B3 to AA10. This data is stored as an array named `a`.
- `[Array,Text,Raw,limits] = xlsread('a.xls', 'hello');`

  The file `a.xls` is read from the worksheet named `hello`, and the numeric data is fed into an array named `Array`. The text data is fed into an array named `Text`. Likewise, the raw cell data is saved into the cell array `Raw` and the ranges are saved as `limits`.
- `xlswrite('new.xls',a)` writes the data in an array named `a` and saves it into an `.xls` formatted Excel sheet named `new.xls`.
- `xlsclose`

```
1  >> pkg load io
2  >> a = rand(10,10);
3  >> odswrite('a.ods',a)
4  ans = 1
5  >> ls
6  a.ods
```

7.3 Accessing Data from the Internet

Very large data sets are often kept on remote servers and you'll need to access them at some point. Using urlread(), you can read remote files. To save data at the local disk, you use the urlwrite() functions.

```
1  >> a = urlread('http://www.fs.fed.us/land/wfas/fdr_obs.dat');
2  >> who
3  Variables in the current scope:
4
5  a    ans
6
7  >> whos
8  Variables in the current scope:
9
10 Attr Name          Size                       Bytes   Class
11 ==== ======        ====                       ======  =====
12 a                  1x147589                   147589  char
13 ans                1x1                        8       double
14
15 Total is 147590 elements using 147597 bytes
16
17 >> urlwrite('http://www.fs.fed.us/land/wfas/fdrobs.dat',
      'fire.dat')
```

```
18  >> ls
19  fire.dat
20  >>
```

Here, a variable named a stores the data from the data file at `http://www.fs.fed.us/land/wfas/fdr_obs.dat`. Alternatively, the whole data file is stored as a file named `a.dat` using the `urlwrite(URL)` function.

7.4 Printing and Saving Plots

Some commands, like `print` and `saveas`, exist to save graphs/figures generated by Octave programs in desired formats. They are discussed in the following sections.

7.4.1 The print Function

The `print` command handles the printing jobs such as printing using a printer and/or plotter, printing to a file, etc. Especially with figures, this command is very useful for saving information automatically with a desired filename in a specified format.

```
1  % Saving in svg format
2  figure(1);
3  clf();
4  peaks();
5  print -dsvg figure1.svg
6
7  % Saving in png format
8  figure(1);
9  clf();
10 sombrero();
```

```
11  print -dpng figure2.png
12
13  % Printing to a HP DeskJet 550C
14  clf();
15  sombrero();
16  print -dcdj550
```

The `clf` function clears the current graphic window. A lot of other options for saving in different formats exist for the `print` command. To learn more, type `help print` at the Octave terminal.

7.4.2 The saveas Function

The `saveas` function saves a graphic object in a desired format, as follows:

```
1  clf();
2  a = sombrero();
3  saveas(a,"figure3.png");
```

7.4.3 The orient Function

The `orient(a,orientation)` function defines the orientation of a graphical object a. The valid options for `orientation` are `portrait`, `landscape`, and `tall`. The `landscape` option changes the orientation so the plot width is larger than the plot height. The `tall` option sets the orientation to `portrait` and fills the page with the plot, while leaving a 0.25 inch border. The `portrait` option (the default) changes the orientation so the plot height is larger than the plot width.

7.5 Summary

This chapter discussed various functions for enabling reading and writing permissions as well as accessing data to and from a file. These actions are an essential part of a numerical computation exercise. Data can be generated in the form of files using software or hardware (an instrument). Octave does not care about its origin. It treats data by its file type. Knowing which function to use to operate on your files is a skill you'll learn and it depends on the situation. File operations allow you to trim data so that only the useful parts of the data are collected. Further trimming can be performed by using slicing operations. By perfecting the art of handling files, you can confidently proceed toward handling sophisticated numerical computations.

CHAPTER 8

Functions and Loops

8.1 Introduction

When a particular numerical tasks needs to be repeated over different data points, digital computers become a useful tool since they can perform repetitive tasks with much greater speed and accuracy than humans. *Loops* perform exactly this task. Using a condition to check the start and termination rules, loops can perform repetitive parts of a process easily. Different programming languages and environments have different rules for defining loops.

Octave provides a much simpler way to define and run loops. They will be discussed shortly. It's useful to define the term *function* here. A big program may require a set of instructions to be called at different times. Hence, these set of instructions can be defined as a subprogram, which can be requested to perform the computation at a desired time. In this way, a complicated task can be divided into many small parts. This architecture of programming is called *modular programming*. This is the most popular way of programming, since it's logical, better at visualizing the problem, and easy to debug. The most popular way of defining these small sets of instructions is to define them as functions. This chapter discusses both of these concepts in detail.

S. Nagar, *Introduction to Octave*, https://doi.org/10.1007/978-1-4842-3201-9_8

8.2 Using Loops

Loops form an essential part of an algorithm since they perform the tasks that computers perform best: doing repetitive actions in a very fast manner. Loops come in many flavors, including the for loop, which repeats certain tasks over a list of variable values, the while loop, which checks for a logical condition before executing a certain task, and the if-then-else loop, which checks a condition and directs the flow of an algorithm. The choice of a particular loop depends on the problem at hand.

A variety of functions and their usage are explained in this chapter. Judging their use critically becomes very important because the looping part of the algorithm consumes most of the execution time.

8.2.1 The while Loop

A while loop defines a logical condition and until that condition is satisfied, it runs a block of code. The syntax for the while loop is as follows:

```
1  while condition
2  BODY
3  endwhile
```

Here, the keyword while initiates the execution of a while loop. The condition is a logical condition whose answer can be true (1) or false (0). The BODY encompasses the commands that are executed until the condition holds true. Listing 8-1 shows an example while loop.

Listing 8-1. The while1.m File

```
1  x = 1.0;
2  while x < 10
3     disp(sqrt(x));
4     x = x+1;
5  endwhile
```

The `while1.m` program runs by first initializing a variable `x` to the value `1.0`. Then it lists a logical condition:

$$x < 10$$

During the first step of the loop, $x = 1$, this condition is satisfied since $1 < 10$. When this condition is satisfied, `disp(sqrt(x))` is executed and displays the square root of `x`. Then line 4 is executed, where `x = x + 1` increments `x`. With a newly incremented value of `x` to 2, the logical condition $x < 10$ is again checked and the body of loop in lines 3 and 4 are executed. This is done until $x = 10$, when the loop condition is no longer satisfied. Then line 5 is executed, which declares the end of the `while` loop. The execution of the `while1.m` file yields:

```
1  >> while1
2  1
3  1.4142
4  1.7321
5  2
6  2.2361
7  2.4495
8  2.6458
9  2.8284
10 3
```

8.2.2 The do-until Loop

It is important to note that there can be cases when the body of a `while` loop is not executed even once. This is the case when, after initialization, a condition is not satisfied. To deal with this kind of scenario, the `do-until` loop has the following syntax:

```
1 do
2 BODY
3 until condition
```

The loop first executes the body of the code and then checks for the condition. This way, the code block comprising the body of the loop is executed at least once. The usage can be understood in the example in Listing 8-2.

Listing 8-2. The dountil1.m File

```
1  %Displaying square root of the
2  %first ten positive natural numbers
3
4  x = 1.0;
5  do
6    disp(sqrt(x));
7    x = x+1;
8  until x == 10
```

The execution of this code yields:

```
 1  >> dountil1
 2  1
 3  1.4142
 4  1.7321
 5  2
 6  2.2361
 7  2.4495
 8  2.6458
 9  2.8284
10  3
11  >>
```

At line 4, x is initialized at `1.0`. Then, the body of the loop displays the square root of x and then increments it by 1. This is done until $x = 10$, i.e., until the value of x becomes `10`.

8.2.3 The for Loop

The for loop is used to perform computations on a list of known values. The syntax of a for loop is as follows:

```
1  for variable = vector
2    BODY
3  end
```

The keyword for declares the start of the loop, where a variable takes the values stored in a vector. Then, the body of the code (represented by BODY) is executed. The keyword end declares the end of the for loop. This is shown in the example in Listing 8-3.

Listing 8-3. The for1.m File

```
1  %program to calculate square root
2  %of the first 10 numbers
3
4  for i = 1:10
5    ans = sqrt(i)
6  end
```

Executing for1.m yields the following:

```
 1  >> for1
 2  ans =  1
 3  ans =  1.4142
 4  ans =  1.7321
 5  ans =  2
 6  ans =  2.2361
 7  ans =  2.4495
 8  ans =  2.6458
 9  ans =  2.8284
10  ans =  3
11  ans =  3.1623
```

8.2.4 The if-elseif-else Loop

In situations where a number of conditions need to be checked at different points in time, the `if-elseif-else` loop works well. The syntax for the loop is given by:

```
1  if condition1
2  BODY1
3  elseif condition2
4  BODY2
5  else
6  BODY3
7  endif
```

On line 1, a condition is defined. If this condition is satisfied, then line 2 is executed; otherwise, line 3 is executed. BODY1 and BODY2 are the blocks of codes that are executed when checking for different sets of conditions, and BODY3 is executed when none of the conditions is executed. Listing 8-4 shows an example of this kind of loop.

Listing 8-4. The ifelse1.m File

```
 1  %Program to check if a
 2  %number is even or odd
 3
 4  x = 33;
 5
 6  if (rem(x,2) == 0)
 7    printf("x is even\n");
 8  elseif (rem(x,5) == 0)
 9    printf("x is odd and divisible by 5\n");
10  else
11    printf("x is odd\n");
12  endif
```

Executing `ifelse1.m` yields:

```
1  >> ifelse1
2  x is odd and divisible by 5
```

At line 4, x is initialized to 33. Then at line 6, the remainder of $\frac{x}{2}$ is checked. If it is zero, then line 7 is executed; otherwise, line 8 is executed, where the remainder of $\frac{x}{5}$ is checked. If it is zero, then line 9 is executed. If both conditions are not satisfied, then line 11 is executed. Line 12 declares the end of the `if-else` loop.

8.3 Using Functions

A *function* is code that can be called as and when required. Hence, it can be defined separately either in a separate file or within the body of the program. Octave presents some ways to define a function, as discussed in the following sections.

8.3.1 The function Function

The definition of a function follows this syntax:

```
1  function [return value 1, return value 2, ...] = name([arg1,
   arg2, ...])
3  body
4  endfunction
```

Here, the `function` keyword defines the object types as a function. Then a set of variables is defined that this function is expected to return. Next comes an equals = operator. Then the name of function. In this case, the function is called `name`. Then comes the main body of the function. The last part defines the end of function.

For example, you can write a function to find x^2 - y^2 and assign it to a variable named z.

```
1  function y = fn1(x,y)
2  y = x^2 - y^2;
3  end
```

Save this as fn1.m in the present working directory. Now go to the Octave terminal and type the following:

```
 1  >> fn1(5,1)
 2  ans = 024
 3  >> fn1(5,2)
 4  ans =  21
 5  >> fn1(5,3)
 6  ans =  16
 7  >> fn1(5,4)
 8  ans =  9
 9  >> fn1(5,5)
10  ans =  0
```

Hence, you can see that the function named fn1 is performing the computation $x^2 - y^2$ on the two input arguments for which it is defined.

It is a good practice to define the program as a group of *function files* and call them in the master program stored as a *script file*. This modular approach makes it easy to experiment with the idea and also makes it easier to debug and test the code. A function can return more than two values too. For example:

```
1  function [y1,y2,y3] = fn2(x,y)
2  y1 = x^2 - y^2;
3  y2 = x^2 + y^2;
4  y3 = y2 - y1;
5  end
```

This gives the following result:

```
1 >> [a,b,c] = fn2(5,2)
2 a =  21
3 b =  29
4 c =  8
5 >> [a,b,c] = fn2(5,0)
6 a =  25
7 b =  25
8 c =  0
```

Functions can incorporate loops to regulate the repetitive tasks inside the program. For example, the factorial of a number can be calculated using the function given here:

```
1 function result = factorial(n)
2   if (n == 0)
3      result = 1;
4      return;
5   else
6      result = prod0(1:n);
7   endif
8 endfunction
```

A function named `factorial`, which takes a number `n` as an argument, calculates the product of the number with all its successive numbers. When called from the Octave command line, the function yields the following result.

```
1 >> factorial(50)
2 ans =    3.0414e+064
3 >> factorial(1)
4 ans =  1
5 >> factorial(0)
```

```
6  ans =  1
7  >> factorial(100)
8  ans =    9.3326e+157
9  >> factorial(1000)
10 ans = NaN
11 >> factorial(-1)
12 error: factorial: N must all be non-negative integers
```

`help NaN` and `help prod` provide useful insights into the behavior of these commands.

8.3.2 The inline Function

Functions can also be defined as *inline* using the `inline` command, as follows:

```
1 >> f = inline("x^2+y");
2 >> f(1,2)
3 ans = 3
4 >> f(10,10)
5 ans =  110
6 >> f(0,2)
7 ans = 2
8 >>
```

Line 1 defines a function named `f` with two variables `x` and `y` to calculate $f(x, y) = x^2 + y$. When called with values of these two variables, the `inline` function outputs the calculated values.

8.3.3 Anonymous Functions

Anonymous functions are unnamed function objects defined in the program. Their definition follows this simple syntax:

```
@(argument list) expression
```

For example:

```
1  >> a = @(x) sin(x)*cos(x);
2  >> quad(a,0,1)
3  ans =  0.35404
4  >> quad(a,0,pi)
5  ans =  7.3031e-017
6  >> quad(a,-pi,pi)
7  ans =  0
8  >> quad(a,-pi,2*pi)
9  ans =  -2.8435e-016
10 >> quad(a,-2*pi,2*pi)
11 ans =  0
```

`help quad` tells us that the function quad evaluated the integration of a function between two values. Hence, line 1 defines a function *sin*(*x*)*cos*(*x*) whose integration is as follows.

$$\int_{0}^{1} sin(x)cos(x) = 0.35404$$

$$\int_{0}^{\pi} sin(x)cos(x) = 7.3031\times10^{-17}$$

$$\int_{-\pi}^{\pi} sin(x)cos(x) = 0$$

$$\int_{-\pi}^{2\pi} sin(x)cos(x) = -2.8435\times10^{-16}$$

$$\int_{-2\pi}^{2\pi} sin(x)cos(x) = 0$$

Hence, using the anonymous function definition, you don't need to name a function.

8.4 Summary

Defining functions is the key to modular programming. Octave presents an elegant way to define and use functions, both inline and in separate files. When combined with the ability to write functions inside a loop, complex problems can be implemented in just a few lines of code. It requires an artistic attitude while designing an algorithm, where functions and loops are the paintbrush to devise an elegant solution to a given numerical problem.

CHAPTER 9

Numerical Computing Formalism

9.1 Introduction

Numerical computation enables you to compute solutions to numerical problems, provided you can frame them into the proper format. This requires certain considerations. For example, if you digitize continuous functions, then you are going to introduce certain errors due to the sampling at a finite frequency. Hence, a very accurate result requires a very fast sampling rate. In cases when a large data set needs to be computed, it becomes computationally intensive and time consuming.

Also, you must understand that the numerical solutions are an approximation at best, compared to analytical solutions. The onus of finding their physical meaning and significance lies on the scientist. The art of discarding solutions that do not have a meaning in a real world scenario is something that scientists/engineers develop over the years. Also, a computational device is only as intelligent as its operator. The law of GIGO (Garbage In Garbage Out) is followed very strictly in this domain.

This chapter explains some of the important steps you must consider to solve a physical problem using numerical computation. Defining a problem in proper terms is just the first step. Making the right model and then using the right method to solve it (solver) is the difference between a naive and an experienced scientist/engineer.

S. Nagar, *Introduction to Octave*, https://doi.org/10.1007/978-1-4842-3201-9_9

9.2 Physical Problems

Everything in our physical world is governed by physical laws. Owing to the men and women of science who toiled under difficult circumstances and came up with fine solutions to things happening around us, we obtained mathematical theories for physical laws. To test these mathematical formalisms of physical laws, we use numerical computations. If they yield the same results as a real experiment, they validate each other.

Numerical simulations can obviate experiments altogether provided you have a well tested mathematical formalism. For example, nuclear powers no longer need to test nuclear bombs because the data about nuclear explosions, which was obtained during actual explosions, enables scientists to model these physical systems quite accurately.

Apart from applications like simulating real experiments, modeling physical problems are good educational exercises. Hands-on modeling exercises enable students to explore the subject in depth and give proper meaning to the topic under study. Solving numerical problems and visualizing the results makes the learning permanent and elucidates any flaws in mathematical theory, which ultimately leads to new discoveries.

9.3 Defining a Model

Modeling means writing equations for a physical system. As the name suggests, an equation is about equating two sides. An equation is written using an equals (=) sign, where the terms on the left side are equal to the terms on the right side. The terms on either side of equations can be numbers or expressions. For example:

$$3x + 4y + 9z = 10$$

This equation has the term $3x + 4y + 9z$ on the left hand side (LHS) and the term 10 on the right hand side (RHS). Note that whereas LHS is an algebraic term, RHS is a number.

Expressions are written using functions, which are simply relationships between two domains. Like $f(x) = y$ is a relationship between y and x using the rules of algebra. Mathematics has a rich library of functions that you can use to make expressions.

The function you choose depends on the problem. Some functions describe some situations best. For example, oscillatory behavior can be described in a reasonable manner using trigonometric functions like $sin(x)$, $cos(x)$, etc. Objects moving in straight lines can be described well using linear equations like $y = mx + c$, where x is their present position, m is constant rate of change of x and y, and c is the offset position. Objects moving in a curved fashion can be described by various non-linear functions (where the power of the dependent variable like x is not 1).

In real life, you can have a mixture of these scenarios. An object can oscillate and move in a curved fashion at the same time. In that case, you write an expression using a mixture of functions or find new functions that could explain the behavior of an object. You verify functions by finding solutions to equations describing the behavior and matching it to observations of the object. If they match perfectly, you obtain a perfect solution. In most cases, an exact solution might be difficult to obtain. In these cases, you'll get an "approximate" solution. If the errors are within tolerable limits, the models can be acceptable.

As discussed, you can analytically solve physical situations by writing mathematical expressions in terms of functions involving dependent variables. The simplest problems have simple functions between dependent variables with a single equation. There can be situations where multiple equations are needed to explain a physical behavior. In case of multiple equations being solved, the theory of the matrix comes in handy.

Suppose these equations define the physical behavior of a system:

$$-x+3y=4 \qquad \text{(Equation 9-1)}$$

$$2x-4y=-3 \qquad \text{(Equation 9-2)}$$

This system of two equations can be represented by a matrix equation as follows:

$$\begin{bmatrix} -1 & 3 \\ 2 & -4 \end{bmatrix} + \begin{bmatrix} x \\ y \end{bmatrix} = \begin{bmatrix} 4 \\ 3 \end{bmatrix}$$

Using matrix algebra, values of variables x and y can be found such that they satisfy the equations. Those values are called the roots of the equations. These roots are the point in 2D space (because there are two dependent variables) where the system will find stability for that physical problem. In this way, you can predict the behavior of the system without actually doing an experiment.

The mathematical concepts of differentiation and integration become very important when you need to work with dynamic systems. When the system is constantly changing the values of dependent variables to produce a scenario, it's important to know the rate of change of these variables. When these variables are independent of each other, you use simple derivatives to define their rate of change. When they are dependent, you use partial derivatives.

For example, Newton's second law of motion says that the rate of change of velocity of an object is directly proportional to the force applied on it. Mathematically:

$$F \, \alpha \, \frac{dy}{dx} \qquad \text{(Equation 9-3)}$$

The proportionality is turned into equality by substituting a constant of multiplication m such that:

$$F = m \times \frac{dy}{dx} \qquad \text{(Equation 9-4)}$$

If you know values or expressions for F, this equation can be solved analytically. In some cases, the analytical solution may be too difficult to obtain. In those cases, you digitize the system and find a numerical solution.

There are many ways to digitize and numerically solve a given function. Programs that implement a particular method to solve a function numerically are called *solvers*. Many solvers exist and the one you choose is critical to successfully obtain a solution. For example, Equation 9-4 is a differential equation. It is a first order ordinary differential equation. A number of solvers exist to solve such equations, including Euler, Runge-Kutta, etc. The choice of the particular solver depends on the accuracy of its solution, the time needed to obtain a solution, and the amount of memory used during the process. Memory usage is important when memory is not a freely expendable commodity, as with micro-computers with limited memory storage.

The advantage of using Octave to perform these numerical computations lies in the fact that it has a very rich library of functions to perform these various tasks. The predefined functions have been optimized for speed and accuracy (in some cases, accuracy can be predefined). This enables the user to rapidly prototype the problem instead of concentrating on writing functions to do basic tasks and optimizing them for speed, accuracy, and memory usage.

9.4 Numerical Approximations

In the course of scientific investigation, finding exact answers may not be possible at times. Instead of devoting a lot of effort to finding an exact answer by solving the problem analytically, another alternative is to develop methods that produce approximate answers. This works well for solutions involving irrational numbers like `pi`. You can choose the number of significant digits for `pi` and determine the accuracy of the result. The degree of accuracy required always depends on the targeted application. For example, when measuring the length of a building, you don't need the answer to be accurate to the length of an atom (Å). Likewise, while measuring a person's body temperature, you don't need it to be accurate to more than two decimal places for most applications. In the era of faster and more efficient computers, you can get higher accuracies by investing more time and storage, whenever required. But this facility must be used judiciously.

9.5 Tolerance

When an approximated answer or a set of approximated answers is available, one of them must be chosen for a particular answer depending on the requirements of the application. One of the ways to make this decision is to define a *tolerance* limit. Tolerance can be defined as a single number or as a range of numbers (having maximum and minimum). The rules that define tolerance limits are application dependent. For example, when measuring human height, you might define the tolerance to be 1 centimeter, whereas when measuring the diameter of a human hair, you would like to be more accurate and measure down to the micron. The decision to define tolerance is simpler when measuring sizes, i.e., tolerance is one or two orders of magnitude smaller than the size of object. It may not be a straightforward task in other applications. For example, measurement of land for constructing a building requires a tolerance of a fraction of meters, whereas positioning a screw in a hole requires the accuracy of fraction of a centimeter.

In mathematical terms, if $\in$ is the tolerance limit, x is the real value and x^* is approximated the value:

$$\left|x - x^*\right| \leq \in \quad \text{(Equation 9-5)}$$

In this case, the absolute error (e_a) and relative error (e_r) of the measurements are given by:

$$e_a = \left|x - x^*\right| \quad \text{(Equation 9-6)}$$

$$e_r = \frac{\left|x - x^*\right|}{x} \quad \text{(Equation 9-7)}$$

Hence, if the absolute error is less than or equal to the tolerance limit, then the approximate solutions are acceptable.

However, if x is known, why do you need to calculate x^*, i.e., an approximate solution?

When solutions of physical systems are unknown, x^* can be calculated and then be compared to the physical measurements. The physical measurements constitute the value of x in this case. By using Equation 9-6, you can calculate any error. Tolerance can then be determined by the fact that some x^* will differ from x *insignificantly,* i.e., the errors won't matter much.

9.6 Taylor Series

Most mathematical functions require many complex operators, other than simpler ones like +, −, x, ÷, to be computed. However a *polynomial* requires only these basic ones to be computed. Hence, if the other mathematical functions can be represented in terms of polynomials, they can be approximated with relative ease.

A polynomial is defined as follows:

$$p(x) = a_0 + a_1 x + a_2 x^2 + \ldots + a_n x^n \qquad \text{(Equation 9-8)}$$

where $a_n \in \Re$ (the a are called the *coefficients*). For the largest n, which corresponds to $a_n \neq 0$, the degree of polynomial is defined to be n.

9.7 Taylor Polynomials

Taylor's theorem shows the way to define a great many mathematical functions, which can be defined as polynomials called *Taylor polynomials.* The accuracy of final answer shown by a Taylor polynomial depends on its number of terms defined in the polynomial. This provides a convenient method to customize the polynomial based on desired tolerance.

Suppose a mathematical function $f(x)$ needs to be approximated around $x = a$. A Taylor polynomial $p_n(x)$ of degree n centered at $x = a$ is the polynomial (of degree at most n) that has the same value as the nth derivative at $x = a$.

Deriving the formula for the Taylor polynomial:

- The zero order polynomial $p_0(x)$ has degree at most 0:
  - $p_0(x)$ must be a constant function (a horizontal line function, graphically)
  - Approximating around $x = a : p(x) = f(a)$
- The first order polynomial $p_0(x)$ has degree at most 1:
  - $p_1(x)$ must satisfy two conditions:

$$p_1(a) = f(a)$$

and

$$p_1'(a) = f'(a)$$

- $p_1(x)$ must be of the form $p_1(x) = mx + c$ (a straight line with slope m and c as the intercept):
- Since $p_1'(a) = f'(a)$ so $m = f'(a)$
- So one can write $c = f(a) - f'(a)a$
- Substituting back values of m and c, you get

$$p_1(x) = f'(a)x + f(a) - f'(a)a = f(a)(x - a)$$

Carrying forward the same arguments in a similar fashion, you can write the general form of the Taylor polynomial of order n as follows:

$$p_n(x) = f(a) + f'(a)(x-a) + \frac{1}{2} f''(a)(x-a)^2 +$$
$$\frac{1}{3!} f'''(a)(x-a)^3 + \ldots + \frac{1}{n!} f^n(a)(x-a)^n$$

This can be rewritten in sigma notation as follows:

$$p_n(x) = \sum_{k=0}^{n} \frac{1}{k!} f^k(a)(x-a)^k \quad \text{(Equation 9-9)}$$

This definition requires that the polynomial must have n derivatives at $x = a$.

The Maclaurin Series is simply the Taylor Series defined for $a = 0$. You can use algebraic manipulations of the Taylor/Maclaurin Series for basic functions like *sin*(x), *cos*(x), e^x, etc. Other complicated functions can also be defined in their series forms. These can be performed by simply using algebraic operators in addition to substitutions, derivatives, and integrations. This mathematical convenience comes in handy when formulating approximate solutions for physical systems defined by complicated functions.

9.7.1 Maclaurin Series for *sin*(*x*) and *cos*(*x*)

To check Maclaurin expansion, let's start with the trigonometric functions *sin*(*x*) and *cos*(*x*). Both are continuous and differentiable in the range given by any set of real numbers. Hence, their differentials exist as well. Thus they can be expanded in the form of a Maclaurin Series as follows.

Suppose $f(x) = sin(x)$ needs to be approximated at $a = 0$. Using Table 9-1 and Equation 9-9 results in:

$$sin(x) = x - \frac{1}{3!}x^3 + \frac{1}{5!}x^5 - \frac{1}{7!}x^7 + \frac{1}{9!}x^9 - \ldots \pm \frac{1}{n!}x^n \quad \text{(Equation 9-10)}$$

Similarly for $f(x) = cos\,(x)$ approximated at $a = 0$.

Table 9-1. *Calculating Coefficients for Maclaurin Series of sin(x) at x = 0*

n	*f*(*x*)	*f(a)*
0	*sin*(*x*)	0
1	*cos*(*x*)	1
0	-*sin*(*x*)	0
1	-*cos*(*x*)	-1
0	*sin*(*x*)	0

Using Table 9-2 and Equation 9-9 results in the following:

$$cos(x) = 1 - \frac{x^2}{2} + \frac{1}{4!}x^4 - \frac{1}{6!}x^6 + \frac{1}{8!}x^8 - \ldots \pm \frac{1}{n!}x^n \quad \text{(Equation 9-11)}$$

Table 9-2. *Calculating Coefficients for Maclaurin Series of cos(x) at x = 0*

n	*f(x)*	*f(a)*
0	*cos*(*x*)	1
1	-*sin*(*x*)	0
0	-*cos*(*x*)	-1
1	*sin*(*x*)	0
0	*cos*(*x*)	1

Choosing Tolerance While Calculating *cos(x)* Using Octave

The `MaclaurinCos.m` file (see Listing 9-1) explains how error is reduced by many orders of magnitude, as more and more terms of Taylor Series are included for calculating cos(15^0).

Listing 9-1. The MaclaurinCos.m File

```
1  %A program to show usage of Taylor Series expansion of cos(x)
2  %Suppose we wish to calculate cos(15) where argument of cos
   function is given in degrees
3
```

```
4  x = 15*pi/180; %converts 15 degrees into radian
5
6  format long %show results in long format having a lot of
                decimal places for numbers
7
8  %Calculating each term of Taylor Series
9
10 p1 = 1;
11 p2 = x^(2)/2;
12 p4 = x^(4)/factorial(4);
13 p6 = x^(6)/factorial(6);
14 p8 = x^(8)/factorial(8);
15 p10 = x^(10)/factorial(10);
16
17 approx_1 = p1-p2; %approximate values using two terms
18 approx_2 = p1-p2+p4; %approximate values using three terms
19 approx_3 = p1-p2+p4-p6; %approximate values using four terms
20 approx_4 = p1-p2+p4-p6+p8; %approximate values using five terms
21 approx_5 = p1-p2+p4-p6+p8-p10; %approximate values using
                                 six terms
22
23 real_value = cos(x); %calculating the real value to find
                          errors
24
25 %calculation of final errors
26
27 error_1 = abs(real_value-approx_1);
28 error_2 = abs(real_value-approx_2);
29 error_3 = abs(real_value-approx_3);
30 error_4 = abs(real_value-approx_4);
```

```
31  error_5 = abs(real_value-approx_5);
32
33  %making an error vector for plotting
34
35  error = [error_1,error_2,error_3,error_4,error_5];
36
37  %plotting error versus number of terms
38
39  figure(1)
40  semilogy(error, '*r-')
41  title('Variation of error in calculating cos(15^{0}) using
    Taylor Series')
42  xlabel('Number of terms on Taylor Series')
43  ylabel('log(error)')
44
45  %plotting cos(x) and its various approximations
46
47  t = 0:0.001:20;
48  %length (t)
49
50  figure(2)
51  y = cos(t);
52  subplot(2,3,1)
53  plot(t,y,t,ones(length(t)))
54  subplot(2,3,2)
55  plot(t,y,t,(1-t.^2/2))
56  subplot(2,3,3)
57  plot(t,y,t,(1-t^2/2+t.^4/factorial(4)))
58  subplot(2,3,4)
59  plot(t,y,t,(1-t.^2/2+t.^4/factorial(4)-t.^6/factorial(6)))
```

```
60  subplot(2,3,5)
61  plot(t,y,t,(1-t.^2/2+t.^4/factorial(4)-t.^6/factorial(6)+t.^8/
    factorial(8)))
62  subplot(2,3,6)
63  plot(t,y,t,(1-t.^2/2+t.^4/factorial(4)-t.^6/factorial(6)+t.^8/
    factorial(8)-t.^10/factorial(10)))
```

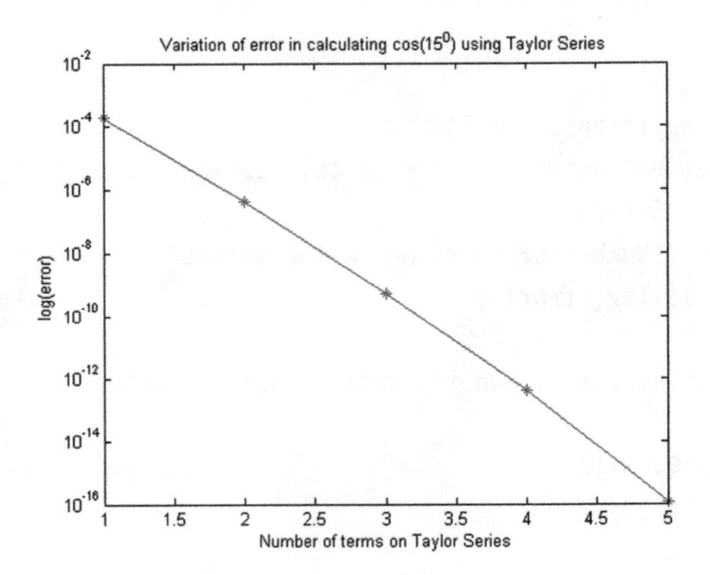

Figure 9-1. *Variation of logarithmic error based on the number of terms used to define a Maclaurin Series for cos(x)*

As you can see in Figure 9-1, you can insert a certain number of terms based on the given tolerance for calculating cos*(x)*. To make a judicious decision about the number of terms, you must inspect the function in a similar fashion (as is done by the code in MaclaurinCos.m). Inserting a lot of terms while demanding less accuracy is a waste of time, energy, and resources (both human and computational).

Instead of expanding around one particular point, the series can be defined for a set of points. The Octave program CosApprox.m shown in Listing 9-2 attempts to do this.

Listing 9-2. The CosApprox.m File

```
 1  %plotting cos(x) and its various approximations
 2
 3  t = -3*pi:pi/10:3*pi;% defining an array of points for x-axis
 4  l = length(t); %to be used for defining pi
 5  y = cos(t); %real values of cosine function
 6
 7  %defining various terms of Maclauren Series
 8  a1 = ones(l); %only first term
 9  a2 = (1-t.^2/2); %first and second term
10  a3 = (a2+t.^4/factorial(4)); %first, second and third term
11  a4 = (a3-t.^6/factorial(6)); %first, second, third and
                                 fourth term
12  a5 = (a4+t.^8/factorial(8)); %first, second, third, fourth
                                 and fifth term
13  a6 = (a5-t.^10/factorial(10)); %first, second, third,
                                   fourth, fifth and sixth term
14
15  %plotting fitting of cos(x) with increasing number of terms
16  figure(1)
17
18  subplot(3,2,1)
19  plot(t,y,'*r-',t,a1,'*b-')
20  axis([-3*pi 3*pi -1.2 1.2])
21  title(' fitting p_{1} to cos(x) ')
22  xlabel(' t ')
23  ylabel(' cos(t) ')
24
25  subplot(3,2,2)
26  plot(t,y,'*r-',t,a2,'*b-')
```

```
27  axis([-3*pi 3*pi -1.2 1.2])
28  title('fitting p-{2} to cos(x)')
29  xlabel('t')
30  ylabel('cos(t)')
31
32  subplot(3,2,3)
33  plot(t,y,'*r-',t,a3,'*b-')
34  axis([-3*pi 3*pi -1.2 1.2])
35  title('fitting p_{3} to cos(x)')
36  xlabel('t')
37  ylabel('cos(t)')
38
39  subplot(3,2,4)
40  plot(t,y,'*r-',t,a4,'*b-')
41  axis([-3*pi 3*pi -1.2 1.2])
42  title('fitting p_{4} to cos(x)')
43  xlabel('t')
44  ylabel('cos(t)')
45
46  subplot(3,2,5)
47  plot(t,y,'*r-',t,a5,'*b-')
48  axis([-3*pi 3*pi -1.2 1.2])
49  title('fitting p_{5} to cos(x)')
50  xlabel('t')
51  ylabel('cos(t)')
52
53  subplot(3,2,6)
54  plot(t,y,'*r-',t,a6,'*b-')
55  axis([-3*pi 3*pi -1.2 1.2])
56  title(' fitting p_{6} to cos(x)')
57  xlabel('t')
58  ylabel('cos(t)')
```

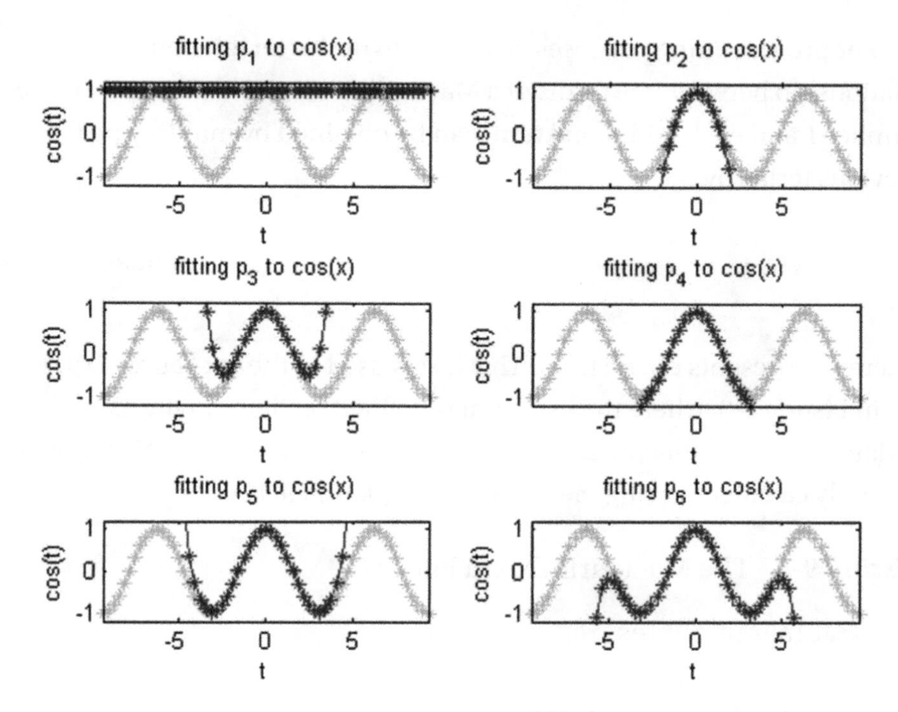

Figure 9-2. *Fitting of Maclaurin Series with a different number of terms to cos(x)*

As you can see in Figure 9-2, as higher order terms are used to describe *cos*(x), the error reduces by fitting with increasing accuracy. For ideal fitting, a very large number of terms must be used to describe the approximated *cos*(x) function. The choice of tolerance is user defined. Depending on the tolerance value, a particular number of terms can be determined.

9.7.2 Maclaurin Series for e^x

Let's explore the concept of errors using another example of Maclaurin Series for e^x:

$$e^a = 1 + a + \frac{a^2}{2!} + \frac{a^3}{3!} + \frac{a^4}{4!} + \ldots \qquad \text{(Equation 9-12)}$$

For programming purposes, it's easier to derive an inherent relationship between the terms of a Maclaurin Series. The first term is the number 1 and each additional term can be obtained by multiplying the previous terms by

$$\frac{a}{n} \qquad \text{(Equation 9-13)}$$

where n represents the nth term. This fact is used in the `MaclaurinExp.m` file in Listing 9-3, where the first term is defined on line 5 in the `expVal` variable and then this variable is added to the variable `currentTerm`, which is simply calculated using the formula in Equation 9-13.

Listing 9-3. The MaclaurinExp.m File

```
1  %Maclaurin Series for exp(0.1)
2
3  n = 5; %Number of terms
4  a = 0.1; %Functional value of x for e^(x)
5  expVal = 1.0;
6  currentTerm = 1.0;
7  for i =1:n
8    currentTerm = currentTerm*a/i;
9    expVal = expVal+currentTerm
10 endfor
11
12 trueVal = exp(0.1);
13 error = abs(trueVal-expVal)
```

The output is displayed as follows:

```
1  >> MaclaurinExp
2  expVal = 1.1000
3  expVal = 1.1050
4  expVal = 1.1052
```

```
5  expVal = 1.1052
6  expVal = 1.1052
7  error = 1.4090e-09
8  >> format long
9  >> MaclaurinExp
10 expVal = 1.10000000000000
11 expVal = 1.10500000000000
12 expVal = 1.10516666666667
13 expVal = 1.10517083333333
14 expVal = 1.10517091666667
15 error =  1.40898115397192e-09
```

Notice that while numeric display is typically set for just four numerical values after the decimal point, the command `format long` increases this setting (the command `format short` returns to the default behavior). You can clearly observe that, by increasing the number of terms, you reduce the error drastically as you approach the true value. You achieve an error in the order of 10^{-9} with just five terms.

If you want to store all the calculated values in the variable `expVal` then you must define it as a vector, as shown in Listing 9-4.

Listing 9-4. MaclaurinExp1.m

```
1  %Maclaurin Series for exp(0.1)
2
3  n = 5; %Number of terms
4  a = 0.1; %Functional value of x for e^(x)
5  expVal = 1.0;
6  currentTerm = 1.0;
7  for i = 1:n
8    currentTerm = currentTerm*a/i;
9    expVal(i+1) = expVal(i) + currentTerm;
10 endfor
11
```

```
12  trueVal = exp(0.1);
13  error = abs(trueVal-expVal)
```

Here, line 9 dictates that the $(i + 1)^{th}$ term is modified based on Equation 9-13 using the previous term, i.e., the $(i)^{th}$ term. Note that printing line 9 has been suppressed here by using the ; operator. The output is shown as follows:

```
1  >> MaclaurinExp1
2  error =
3
4  1.0517e-01    5.1709e-03    1.7092e-04    4.2514e-06
   8.4742e-08    1.4090e-09
5  >>>plot(error, 'r*-')
```

Using `plot(error,'r*-')`, you can generate graphs of error values. This is illustrated in Figure 9-3.

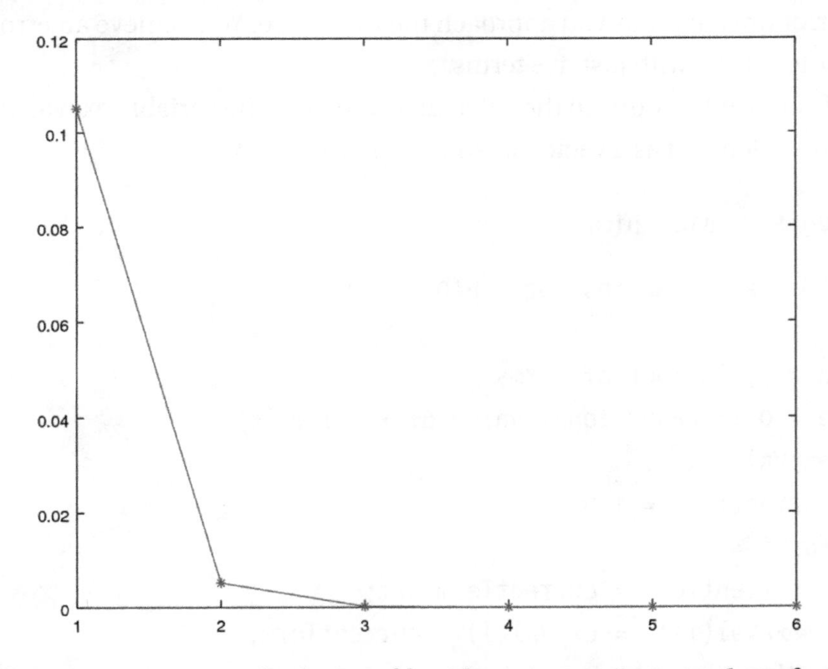

Figure 9-3. *Error in calculating $e^{0.1}$ with an increasing number of terms*

Since the error drops by orders of magnitude with each new term, the effect can be best seen using a logarithmic plot. This can be generated using the `semilogy(error,'r*-')` command; Figure 9-4 will be generated.

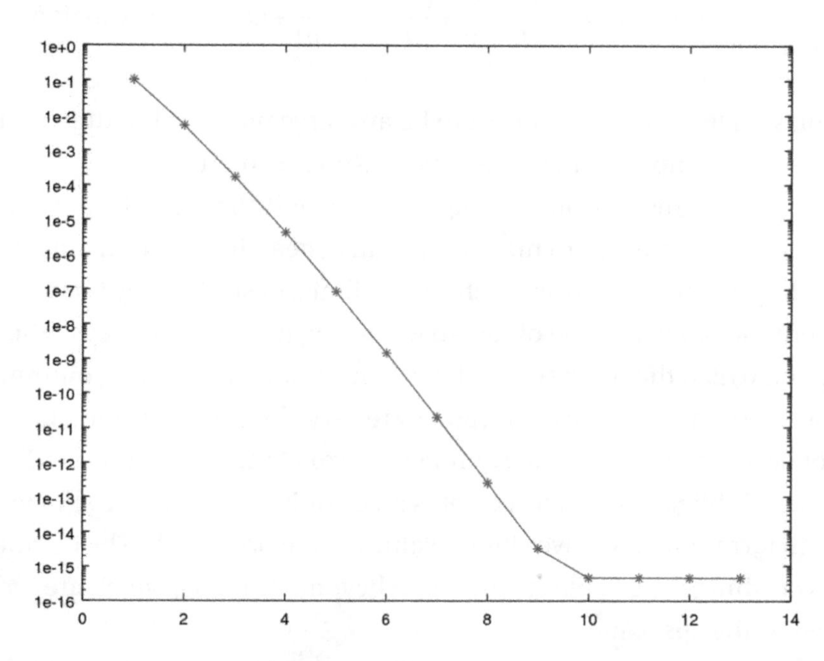

Figure 9-4. *Error in calculating $e^{0.1}$ with an increasing number of terms*

It seems that beyond 10 terms, the error flattens out. But you shall see that this is an erroneous result, as this graph will depend on the least count of the computing machine.

How Many Terms

You can observe from Figure 9-4 that by increasing the number of terms, you reduce the error by two orders of magnitude when calculating $e^{0.1}$. But does this trend mean that to achieve true values, you must include *1* number of terms? After all, each time you add a new term, you invest time and energy resources into the computation. In general, the Maclaurin Series has the accuracy of a^{n+1} when n terms are used:

$$e^a = 1 + a + \frac{a^2}{2!} + \frac{a^3}{3!} + \frac{a^4}{4!} + \ldots + \frac{a^n}{n!} + O\left(a^{n+1}\right) \quad \text{(Equation 9-14)}$$

Analytically, you can choose n to be any large number, but this cannot be done on a computing machine. The reason is explored next.

Figure 9-4 shows one interesting fact—beyond 10 terms, the error no longer changes by orders of magnitude and instead flattens out. This is misleading. Each computing machine has limits of storing tiny floating point numbers. This can be obtained by issuing the command `eps`. The system on which the program has been run shows the following output. You can see that when the error values are very close to `eps` values, they cannot be stored reliably. The command `error(9:12)` outputs a similar viewpoint. While the ninth term yields an error in the order of 10^{-15}, the eleventh terms onward have similar values in the order 10^{-16}. The computer avoids crashing the calculation by going beyond its limits, which are defined by the eps value.

```
1  >> error(9:12)
2  ans =
3
4  3.10862446895044e-15   4.44089209850063e-16
   4.44089209850063e-16   4.44089209850063e-16
5  >>> eps
6  ans =  2.22044604925031e-16
```

The command `eps` gives the machine precision. The command `help('eps')` shows the documentation for the `eps` command and its usage. Technically, `eps` is the relative spacing between any two adjacent numbers in the machine's floating point system, i.e., these computational machines *least count*. This number is obviously system dependent, as you could devise specialized hardware where machine precision can be enhanced. In fact, this is done for cases where increased precision really matters, such as for missile guidance, space navigation, etc. On machines that support IEEE floating point arithmetic, `eps` is approximately 2.2204 x 10^{-16} for double precision and 1.1921×10^{-7} for single precision.

It is interesting to know that $2^{-52} \approx 2.2204\times10^{-16}$ essentially signifies that the *double precision* mode of the software can store 52 digits after the decimal point.

The `realmax`, `realmin`, `intmax`, and `intmin` commands show the maximum and minimum values of real numbers and integers on the particular machine where the software is installed.

```
1  >> realmax
2  ans =  1.79769313486232e+308
3  >> intmax
4  ans =  2147483647
5  >> realmin
6  ans =  2.22507385850720e-308
7  >> intmin
8  ans =  -2147483648
```

It is useful to know these numbers, as any numbers beyond these limits will be prone to error due to machine precision.

9.8 Computational Error

You have learned about the inherent errors due to the inclusion of certain numbers of terms while calculating a mathematical function. There is another kind of error as well, which is introduced due to the fact that computers can store numbers of finite lengths.

9.8.1 Significant Digits

The concept of significant digits plays an important role here. If computers can store all the significant digits of the final answer, then the errors become irrelevant. Otherwise, it is important to identify them and if possible rectify them when reporting a final solution. For example, while dealing with `pi`, if only three significant digits are desired, it can be stored easily on any low-end computing solution.

Computers can store numbers as floating point objects. A floating point object stores a number as follows:

$$\pm d_1 d_2 \ldots d_s \times \beta^e \qquad \text{(Equation 9-15)}$$

Where $d_i = 0,1,2 \ldots \beta - 1$ but $d_1 \neq 0$ and $m \leq e \leq M$ where $m \in I^-$ and $M \in I^+$.

The three parts of a floating point number are as follows:

- Sign $(\pm)$
- Mantissa $(d_1 d_2 \ldots d_s)$
- Exponent (β)

Each part is stored in its own separate fixed-width storage space. Based on the IEEE double precision roundoff, MATLAB uses *binary arithmetic*, whereby:

- $\beta = 2$
- $s = 53$
- $m = -1074$
- $M = +1023$

Since humans are used to *decimal arithmetic* systems, these binary numbers are converted to decimal numbers for reporting purposes. It is important to understand this key point—all internal calculations are done in binary form but input and output is in decimal form. The rounding-off error due to these conversions is given by the *unit roundoff u*, which is the maximum relative error while approximating a real number as a floating point number.

MATLAB can handle numbers with absolute values from $2^{-1074} \simeq 10^{-324}$ and $2^{-1023} \simeq 10^{308}$ with a unit roundoff $u = 2^{-53} \simeq 10^{-16}$.

9.9 Challenges in Real Number to Floating Point Number Conversions

A real number *x* can be stored in a floating point representation given by Equation 9-15 as follows:

$$x = \pm d_1 d_2 \ldots d_s d_{s+1} \ldots \times 10^e \qquad \text{(Equation 9-16)}$$

Note that *s* = 53, but the previous description does not limit the representation of the floating point number. Its storage is an altogether different story. When it is stored, the number is rounded off and stored based on the guidelines—*s* = 53 in this case.

9.9.1 Overflow

From Equations 9-15 and 9-16, if $e > M$, computation is said to have *overflowed,* i.e., a number bigger than the possible storage has been presented and hence the storage container has *overflowed.* In this case, MATLAB produces `Inf` or `-Inf` as the answer, which represents the fact that the answer is a very large number. The following exercise, performed in a MATLAB terminal, illustrates the process clearly. `Inf` is displayed as an answer when e^{900} is attempted. When this number is divided by a negative number, `-Inf` is displayed, signifying an overflow while storing a negative number. When `Inf-Inf` is attempted, NaN is displayed, signifying that the large numbers cannot produce a meaningful result.

```
1  >> format long
2  >> exp(50)
3
4  ans =
5
6  5.184705528587072e+21
7
8  >> exp(100)
9
10 ans =
11
12 2.688117141816136e+43
13
14 >> exp(500)
15
16 ans =
17
18 1.403592217852837e+217
19
20 >> exp(700)
21
22 ans =
23
24 1.014232054735005e+304
25
26 >> exp(900)
27
28 ans =
29
30 Inf
31
```

```
32  >> exp(900)/-2
33
34  ans =
35
36  -Inf
37
38  >> exp(900)-exp(900)
39
40  ans =
41
42  NaN
```

9.9.2 Underflow

If $e < m$, then *underflow* has occurred. Octave represents underflow by showing zero as an answer. You might think that that underflow is not serious, but consider the fact that, based on the basic rules of exponentiation:

$$e^a e^{-a} = e^{a-a} = e^0 = 1$$

When you perform the same calculations for numbers representing overflow and underflow, Octave has to perform `Inf X 0`, which results in `NaN`. This is demonstrated in this example:

```
1  >> exp(900)*exp(-900)
2
3  ans =
4
5  NaN
6
7  >> exp(900)
8
```

```
 9  ans =
10
11  Inf
12
13  >> exp(-900)
14
15  ans =
16
17  0
```

9.10 Converting Real Numbers to Floating Point Numbers

After understanding the two extreme cases, overflow and underflow, you need to understand the process of real number to floating point number conversion. Recall from Equations 9-15 and 9-16 that a real number can be stored with *s* significant digits, as follows:

$$\pm d_1 d_2 \ldots d_s \times \beta^e$$

This can be written in floating point notation (for base-10) as:

$$x = \pm d_1 d_2 \ldots d_s d_{s+1} \ldots \times 10^e$$

There are two ways to achieve the conversion: *method of truncation* and *method of rounding off*. Method of truncation will simply discard all digits after *s*, i.e., it will produce:

$$x = \pm d_1 d_2 \ldots d_s \times 10^e \qquad \text{(Equation 9-17)}$$

On the other hand, the method of rounding off recommends the following process:

1. If $s_{s+1} < 5$ then perform truncation and retain the sign of x.
2. If $s_{s+1} > 5$ then d_s is incremented and the truncation is performed. Retain the sign of x.

This seemingly simple scheme has a flaw. Suppose for $s = 4$, you need to round off 2.9345. The answer is 2.934, i.e., the last digit is 5 and it's simply discarded. In a similar fashion when 2.9355 is rounded off, the answer can be written as 2.936, where the last digit is discarded and last-significant-digit is incremented. In both cases, only one digit needs to change. But suppose you need to round to 2.9999. In this case, the answer comes out to be 3.000, where four numeral values need to be changed.

9.11 Octave Packages

A number of packages exist to perform numerical computations in a particular scientific domain. The reference [1] lists some of these packages. You can install a package using this command on the Octave command line:

```
>> pkg install -forge package_name
```

9.12 Summary

Almost all branches of science and engineering require you to perform numerical computation. Octave is one of the alternatives to doing so. Octave has a library of optimized functions for general computation. Also, it has a variety of packages to perform a specialized job. This makes it an ideal choice for prototyping a numerical computation problem efficiently. This chapter summarized various issues related to the error generated during numerical computation and various methods to obtain their value or order of magnitude. These quantities are important to measure since in real life, you will need these values to define the accuracy of the final product.

This book presented the Octave programming language as an effective alternative to the MATLAB base package. Furthermore, additional packages can be associated with the Octave framework to perform calculations from a specific domain. With an active community of developers, Octave is flourishing in industry and academia and definitely has a bright future.

9.13 Bibliography

[1] https://octave.sourceforge.io/

Index

A, B

S. Nagar, *Introduction to Octave*, https://doi.org/10.1007/978-1-4842-3201-9

C, D, E

F

G

H

I, J, K

L

M

T

U, V, W, X, Y, Z

CPSIA information can be obtained
at www.ICGtesting.com
Printed in the USA
LVHW011633200723
752868LV00001B/8